Implementing e-Navigation

For a listing of recent titles in the
Artech House GNSS Library,
turn to the back of this book.

Implementing e-Navigation

John Erik Hagen

ARTECH
HOUSE

BOSTON | LONDON
artechhouse.com

Library of Congress Cataloging-in-Publication Data
A catalog record for this book is available from the U.S. Library of Congress.

British Library Cataloguing in Publication Data
A catalog record for this book is available from the British Library.

ISBN-13: 978-1-63081-099-3

Cover design by John Gomes

10 9 8 7 6 5 4 3 2 1

Contents

Acknowledgments

I wish to thank various people for their contribution to this book. Special thanks should be given to Jon Leon Ervik and Michael Rambaut for their professional guidance and constructive recommendations and inputs on this project.

I would also like to acknowledge Aileen Storry, my editor, assisted by Holly Smith-Young and others at Artech House. Aileen initiated the project and has, together with several people at Artech House, provided valuable and relevant inputs during the entire process. I also appreciate Kirsti L. Slotsvik for support of the project idea and encouragement for keep up with this task.

This book is an output of combined effort of many people. I would like to express my very great appreciation to those who contributed in some way to the work or agreed to be interviewed: Gilles Bessero, Francis Zacharie, Nick Lemon, Mahesh Alimchandani, Michael Bergmann, Lene Vesterlund, Frances Baskerville, Kwan An, Javier Yasnikouski, Lawrence Sciberras, Brian J. Tetreault, Hiroyuki Yamada, Dmitry Rostopshin, Anne Grethe Nilsen, David Patraiko and the Nautical Institute, Pieter Paap, Kevin Richardson, Finn Martin Vallersnes, Alexander McDonald, Jerzy Najmola, Florian Motz, Johan Gahnstrom, Pierre D'Arcy, Jan Fredrik Meling, Jorge Pecci, Tormod Våga, Simon Pelletier, Lasse Karlsen, and Seojeong Lee. Without their participation and input, the book could not have been successfully completed.

Last and foremost, I am dearly grateful to my wife, Jane, for her patience and support. This accomplishment would not have been possible without her. Thank you.

1

Introduction to e-Navigation

1.1 What Is e-Navigation?

In 2016, the new secretary-general of the International Maritime Organization (IMO), Mr. Kitack Lim, delivered the keynote speech at the conference e-Navigation Underway onboard the M/S *Pearl Seaways* in Copenhagen, Denmark. This was the first major external event he had been invited to address in his new role as secretary-general of IMO.

"One of my most important objectives at IMO is to ensure that the Organization continues to draw the experience and the knowledge of those involved in shipping at the operational level, so being here on a ship has a symbolic resonance, in that respect, that I really appreciate," he said, and thereafter stated: "e-Navigation is the future; but it has been 'the future' for a long time." He further explained his statement: "The challenge now is to turn 'the future' into 'the present' so that all the much heralded benefits and advantages of e-Navigation can be fully realized" [1].

A formal definition of e-Navigation was adopted in 2008: "e-Navigation is the harmonized collection, integration, exchange, presentation and analysis of maritime information onboard and ashore by electronic means to enhance berth to berth navigation and related services, for safety and security at sea and protection of the marine environment." The need for e-Navigation is embedded in the IMOs e-Navigation strategy plan for the development and implementation of e-Navigation, which included approval of the definition [2].

The result of further development of the IMO's strategy plan was the finalization of the e-Navigation Strategy Implementation Plan (SIP), which is based on five prioritized e-Navigation solutions. This plan was approved after a comprehensive process six years later, in 2014. The SIP is mainly a guide for all stakeholders to raise awareness of specific user needs related to e-Navigation and prioritized solutions to address these user needs in a cost-effective manner. This work also has to be done in a harmonized and standardized manner.

The e-Navigation development process involved many stakeholders, such as shipborne and shore-based users, including IMO member states as flag, coastal and port states, and international organizations, industrial actors, research institutes, and service providers.

The strategy plan reflects expectations of the IMO as well as other stakeholders. It is at the same time also interesting to look at examples of e-Navigation challenges and already existing operational best practices based on the e-Navigation philosophy. However, moving rapidly in the digital age, the reader should take into account the fact that information and data found here could inevitably and quickly become out of date.

The IMO Strategy Plan outlines that "the aim of e-Navigation is to meet present and future needs through the harmonization of marine navigation systems and supporting services. The overall goal is to improve safety of navigation and to reduce errors by equipping users, on ship and ashore, with modern, proven tools, optimized for good decision-making, to make maritime navigation and communication more reliable, and user-friendly" [2].

The strategy plan states that the IMO is the one single institution having the technical, operational and legal competences needed to define and enforce the overarching framework for the implementation of e-Navigation.

In his speech in Copenhagen in 2016, the secretary-general underlined, however, that the IMO did not have to carry out all the relevant tasks in-house. "There are other stakeholders which are quite properly taking the lead in many different areas, according to their competence and expertise," he said, and emphasized that the development of e-Navigation is clearly a collective task where the IMO plays a central and coordinating role.

The secretary-general also brought focus on the "e" in e-Navigation, which can be used in several connotations: "Does it stand for 'electronic' navigation?" he questioned and acknowledged that electronic navigation had existed for a very long time; however, "e-Navigation also looks at issues that are clearly not electronic, such as operational procedures, familiarization, documentation and manuals and, of course, training."

Mr. Kitack Lim continued: "Perhaps the 'e' stands for 'enhanced' navigation—there is certainly a clear understanding that e-Navigation should deliver an improvement on what we already have. Of course, back in 2008, when the first proposal on this subject was presented to IMO, 'e' was the fashionable prefix to indicate something advanced and ground-breaking; maybe if that proposal was initiated today, it would talk about 'i-navigation'—'i' for integrated, or perhaps 'i' for improved. But what we have is e-Navigation: and I think the most appropriate way to define this might actually be 'evolved' navigation," the secretary-general concluded [1].

1.2 The Vision of e-Navigation

e-Navigation developments in communication and information technology provide opportunities to develop knowledge management in order to increase transparency and accessibility to information. The challenge is to make sure the vision and the strategy for e-Navigation allows that evolving potential to be fully realized.

The vision of e-Navigation is embedded in general expectations for onboard, ashore, and communication elements (see Figure 1.1). In other words; the onboard element is about navigation systems that benefit from the integration of ship sensors, supporting information, a standard user interface, and a comprehensive system for managing guard zones and alerts. Core elements of such a system

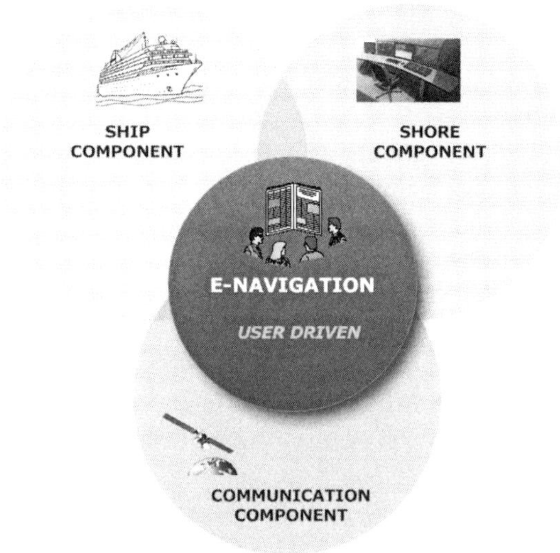

Figure 1.1 Key components of e-Navigation as presented by the Norwegian Coastal Administration. (*Source:* Norwegian Coastal Administration.)

will include active engagement of the mariner in the process of navigation to carry out the navigator's duties in the most efficient manner, while preventing distraction and overburdening.

The focus ashore is on the management of ship traffic and related services enhanced through better provision, coordination, and exchange of comprehensive data in formats that will be more easily understood and utilized by shore-based operators in support of vessel safety and efficiency.

Last, communications is about providing an infrastructure to facilitate authorized seamless information transfer onboard ships, between ships and shore, and between authorities and other parties with many related benefits [3].

1.3 Development of e-Navigation

1.3.1 The e-Navigation Concept

e-Navigation digitizes and standardizes the data and information used for safe and efficient navigation and provides decision-making support for navigators on board ships. The concept also reduces the administrative workload on board, so that navigators can focus on tasks related to safety of navigation and efficient maritime operations. e-Navigation harmonizes and simplifies digital information exchange between systems on board, systems on shore, and systems between other ships. It involves the integration of new and existing bridge technologies and equipment to enable the provision of globally harmonized maritime services. It facilitates a simplified and harmonized bridge design, ensures that relevant navigation information is displayed appropriately, and that navigation equipment can be operated in a standardized manner. e-Navigation requires communications infrastructure for digital transmission of integrated and once-only ship reporting to authorities and digital services provided from shore-based authorities and other stakeholders to ships. Risk analysis based on IMO Formal Safety Assessment methodology shows that e-Navigation could reduce the risk of accidents, both collisions and groundings.

The e-Navigation concept is based on five solutions prioritized by the IMO. It represents an integrated and holistic approach to the concept of exchange of information in providing safe, secure, and effective berth-to-berth navigation as well as protection of the marine environment. Some of the solutions are within the scope of the IMO and some of them are within the scope of other stakeholders such as organizations or industry. What is important is that the development was done in a coordinated manner.

The first solution, S1, is improved, harmonized, and user-friendly bridge design. It includes functionality to the Integrated Navigation System (INS) performance standard, which will facilitate a simplified and harmonized bridge design and ensure that relevant navigation information is displayed appropriately. In addition, the solution outlines a standardized mode of operation (S-mode) and display for all navigational equipment, and provides seafarers with the ability to operate navigation equipment in a standardized manner, thereby improving the safety and efficiency of navigation.

The means for standardized and automated reporting is the second solution, S2. The benefit will be a worldwide harmonized and, where possible, automated, ship reporting system reducing cost and administrative burden, while also reducing the nonnavigational workload of a navigator and increasing the efficiency of trade.

The third solution, S3, is improved reliability, resilience, and integrity of bridge equipment and navigation information. There is a compelling need to ensure that anavigator is presented with information that is accurate and reliable at all times with an indication when the equipment is not working satisfactorily. Requirements for built-in integrity testing of navigation equipment will ensure navigators are confident that information presented to them is accurate.

Integration and presentation of available information in graphical displays received via communication equipment is covered in solution 4, S4. The display of information on board in a harmonized, effective, and user-friendly way increases the overall awareness of data, which in turn improves situational awareness and reduces mistakes made in transferring information from paper outputs.

Solution 5, S5, is the framework for the electronic provision of information related to maritime services in a harmonized way between shore and ships, which are called maritime service portfolios (MSPs). This solution, originally called improved communication of vessel traffic service (VTS) portfolio; however, is not limited to VTS and aims to harmonize the format, structure, and communication channels used to exchange that information. Sixteen services or MSPs have been identified in the SIP.

Solutions S2 on standardized and automated ship reporting, S4 on integration and presentation of available information in graphical displays, and S5 on MSPs focus on efficient transfer of marine information and data between users, such as ship-to-ship, ship-to-shore, shore-to-ship, and shore-to-shore. Solutions S1 on improved, harmonized, and user-friendly bridge design, and S3 on improved reliability, resilience, and integrity of bridge design and navigation information promote workable and practical use of the information and data on board.

Similar to e-Navigation development, the Marine Electronic Highway (MEH) was a Demonstration Project of the IMO in the Straits of Malacca and Singapore, which was given the go-ahead in June 2006. The overall objectives of the project were to enhance maritime services, improve navigational safety and security, and promote marine environment protection. The four-year regional demonstration project aimed to link shore-based marine information and communication infrastructure with corresponding navigational and communication facilities aboard transiting ships, while being also capable of incorporating marine environmental management systems. In general, if e-Navigation solutions are realized as intended, then the concept of a MEH on a global basis can be achieved. The entire e-Navigation concept is illustrated in Figure 1.2

1.3.2 e-Navigation Is a Collective Task

There is no doubt that the development of e-Navigation is a collective task where although the IMO plays a central and co-coordinating role, other stakeholders, such as the International Association of Marine Aids to Navigation and Lighthouse Authorities (IALA) and the International Hydrographic Organization (IHO) have taken the lead in different areas according to their competence and expertise. Later in this chapter there is also a discussion about the industry's role in the development.

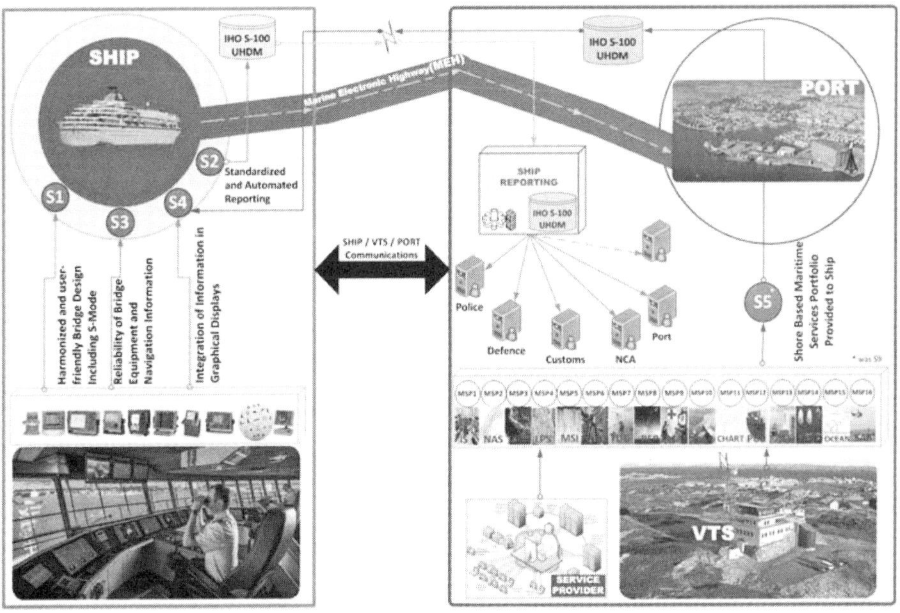

Figure 1.2 The e-Navigation concept. (Illustration: Ulrich Tagne.)

The IALA has dedicated a committee to work on different aspects of e-Navigation, such as data modeling and message systems, e-Navigation communications, shore technical infrastructure, e-Navigation test beds, and MSPs related to the IALA scope. These aspects aim to define and harmonize the structure of MSPs and to provide guidance on practical implementation of MSPs and the appropriate communication channels used for the electronic exchange of information between shore and ships.

The IHO develops the specifications of MSPs related to its scope and contributes to the development of a harmonized display of navigation information through S-100—the IHO Universal Hydrographic Data Model—and associated Product Specifications. Several Product Specifications based on S-100 will form an integral part of e-Navigation and are being developed by the IHO and other organizations, such as the Joint Technical Commission for Oceanography and Marine Meteorology (JCOMM), a joint body of the Intergovernmental Oceanographic Commission (IOC) and, the World Meteorological Organization (WMO) as well as IALA.

Ensuring the harmonized display of information related to more than one navigational data product is part of the broader issue of interoperability between the relevant Product Specifications. The IHO also investigates whether the use of S-101-based electronic navigational charts (ENCs) and possible simultaneous display of other data products would require any amendment to the electronic chart display and information system (ECDIS) performance standard. In addition, the International Electrotechnical Commission (IEC) contributes to the preparation of standards for maritime navigation and communication equipment and systems, and reports to the IMO on the progress of work that is relevant to e-Navigation. More about standards is described in Chapter 5.

1.3.3 Approaches toward a Global e-Navigation System

On a global level, e-Navigation will standardize bridge design that globally enhances the opportunity to work cross-border, improves efficiency in training, and reduces material cost. Similarities between nations and vessels will also increase efficiency and improve safety, reduce barriers of trade through reduction of local solutions and bureaucracy, and reduce the risk of accidents and incidents.

For coastal states, flag states, and port states, e-Navigation will improve situational awareness by providing easy access to standardized and reliable information and improve efficiency in supervision, coordination, and control, as well as coordination and information. Furthermore, e-Navigation will reduce the risk of accidents and incidents through efficient use of shore-based services provided to ships from VTSs, ports and other stakeholders.

However, the idea behind e-Navigation is that it can only be made to work as an international—indeed, global—system. The overall aim is to achieve worldwide implementation. But there is also much being done at national and regional levels in this respect. Below are some examples from Asia and Europe.

In order to enhance maritime safety, the government of the Republic of Korea (ROK) set up a national strategy plan for development and implementation of e-Navigation in 2013, and finalized a national strategy implementation plan in 2015.

The main benefits of e-Navigation are improved safety and better protection of the marine environment through the promotion of navigational system standards and a reduction in human error. According to statistics in the ROK, provided by Mr. Kwan An, former Deputy director, Ministry of Oceans and Fisheries [4], most marine accidents involved fishing vessels and smaller ships of under 100 gross tons (GT): 88.4% of the ROK's registered ships are fishing vessels and 94.9% of registered ships are smaller ships under 100 GT. Recognizing the importance of the safety of non-Safety of Life at Sea (SOLAS) ships, the national e-Navigation strategy focuses on non-SOLAS ships, such as coastal operating ships and fishing vessels in the coastal waters of the ROK.

The ROK's SIP provides a national framework for the development and implementation of e-Navigation. According to their SIP, the national e-Navigation project will be performed from 2016 to 2020 with approximately 114 million USD of government budget. The project has three strategies: (1) development of core technologies of e-Navigation services, (2) expansion of infrastructure, and (3) development of e-Navigation standards based on IMO's SIP.

First, the core technologies of e-Navigation services include technology for situational awareness of maritime traffic and response to navigational hazard based on real-time traffic monitoring system together with ship position information. It also includes technologies for risk assessment for identified navigational hazards. The developed technologies will be applied to e-Navigation services for non-SOLAS ships.

Second, a communications system and an e-Navigation operations system will be established as an e-Navigation infrastructure. A communications system is key for the implementation of e-Navigation.

Last, the development of standards or guidelines for software quality assurance (SQA), standardized mode for navigational equipment (S-mode), Maritime Cloud, and S-100 Data Model will be performed under the e-Navigation SIP in the ROK.

Identified e-Navigation services for non-SOLAS ships under the national e-Navigation SIP are as follows[4]:

- ENC download and ENC up-date service;
- ENC streaming service;
- Vessel monitoring and remote support service;
- Routing support service;
- Communication support service between SOLAS ships non-SOLAS ships

The European Union (EU) regional project Accessibility for Shipping, Efficiency Advantages and Sustainability (ACCSEAS) "identified key areas of shipping congestion and limitation of access to ports, [and] defined solutions by prototyping and demonstrating success in an e-Navigation test-bed at North Sea regional level." The project was supported by six countries in the region. ACCSEAS was a phased program to recognize problem areas and determine solutions using the IMO e-Navigation concept and European initiatives [5].

The project supported "improved maritime access to the North Sea Region through minimizing navigational risk. With European transport policy providing a shift to seaborne transport, using short sea shipping to avoid road bottlenecks to the movement of goods, services, and people, efficient and effective marine navigation services have never been more important.

By looking to harmonize maritime information and how it is exchanged and by offering training provision to support real-world implementation, ACCSEAS has worked to ensure that e-Navigation provision in the North Sea contributes a beneficial and lasting impact on the resilience of the region's critical infrastructure in terms of safety, security, economic growth, and environmental protection.

ACCSEAS has built on the findings of previous and current related regional projects and focused on cooperation in key areas of technology and infrastructure services that underpin maritime navigation and safety—looking to further enhance them" [5]. The project has followed the guidance and regulatory framework of the IMO, EU, and IALA, and has looked to prove the success of new e-Navigation concepts by producing key types of results and outputs.

These outputs have and hopefully continue to "inform policy development and influence the creation of any necessary institutional structures and regulatory instruments needed to deliver future e-Navigation aids to navigation services" [5].

These two examples demonstrate the adherence to the IMO e-Navigation concept, showing examples on how to implement and harmonize national and regional solutions as a contribution to a global implementation of e-Navigation.

1.3.4 Industry's Role

Comité International Radio-Maritime (CIRM) is an international association made up of many different types of companies but all are involved in some way or another with the design, manufacture, delivery, installation, and service of electronic navigation and communication equipment.

A technical program such as e-Navigation, which will add new features and applications to an existing navigation and communications system such as those found on ships today, can only progress with the support of the designers and manufacturers of such systems. For convenience, the IMO deals with only one accredited body to represent the interests of manufacturers of electronic equipment onboard, this being CIRM. This organization is one of many nongovernmental international organizations in consultative status used by the IMO when developing new systems.

Manufacturers are important since they will be the ones who implement any changes by providing updates or new equipment when it is clear what is needed. They have the technical expertise necessary to show whether new ideas are feasible with the technology available. They can also develop test beds to prove ideas and use the information to guide the IMO in developing performance standards. CIRM provides manufacturers with a unified voice at the IMO.

It is important to note that the manufacturing industry is driven by profit from developing and producing products. In the case of e-Navigation, there is a difficulty in justifying a program to develop something new unless there is an opportunity for sales. Most of the current ship fit is regulated by carriage requirements. Without some sort of implementation date and carriage requirement, it could be difficult to plan a rollout date for any new products and reach a return on any investment made in development of such products.

For existing systems, equipment, and interfaces it is of vital importance that there is harmonization, including data formats of the symbols used. This work requires a coordinated approach between international organizations and the industry, which have important roles in the design and development of equipment and systems.

However, for branches, organizations, and industry, e-Navigation will eventually provide flexibility with regard to training and familiarization since standardization would lead to a more efficient market for standardized bridge products and provide a direction for product development to a wide market and opportunity for new products and solutions.

In personal conversation, Ms. Frances Baskerville, secretary-general of CIRM said, "The manufacturers have been vocal enough and said that the entire e-Navigation process to date, has taken far too long. It's a challenge for industrial and commercial participants to convince their management to stay engaged in the process. But now they do have the test beds to hook into. Manufacturers and industrial partners also need a clear timeline for when they can get approvals, and start to place equipment and services on the market. A global regulatory requirement and nationally encouraged usage, mandates or mandatory drivers from the shipping and port communities will, needless to say, provide gains in efficiency and again, ultimate cost reductions."

So, can e-Navigation systems and equipment represent business opportunities for the future?

In 2015, Frances Baskerville asked for opinions on what CIRM members thought e-Navigation was—how did they define it? The replies were varied and interesting and she kindly shared them here. Of course, most looked at what e-Navigation meant to them from their own companies' point of view rather than taking a holistic approach.

One CIRM member, a search and rescue company, stated: "Our impression is that e-Navigation is mostly concerned with automating all data processes between ship and shore in order to streamline safe navigation and trade. If we get involved with it, it will be part of the whole 'Internet of Things' whereby every widget will eventually be 'connected.' Maybe even emergency position indicating radio beacons (EPIRBs) will regularly test themselves and let us know when they need a new battery!"

Another said: "e-Navigation allows higher control of the vessel security wise, of onboard system performance, engines, navigation, communications and automation systems. e-Navigation allows remote assistance. It is an emerging way of providing a maintenance service but that maintenance service needs to be regulated for security reasons. Protocols and different technologies need to be aligned in order to make this possible, and efficient. Remote assistance and configuration saves time. In other words, everyone saves time and money."

In the security-conscious Middle East, a CIRM member said: "The existence onboard of a dedicated recording system voyage data recorder (VDR), and the recording of data on other systems such as ECDIS, has great benefit for auditing. Masters, owners, and authorities can replay and analyze the performance of the ship and officers in relation to several safety aspects such as collision avoidance, risk of grounding and possibly others. e-Navigation should give more emphasis to the ability to analyze such data and inform clearly when a 'near miss' or 'near grounding' happened, or when a dangerous action (or inaction) was taken. The results should assist in auditing but would also be very important in embedding and implementing standards of navigation safety and safety education."

One CIRM company heavily involved in defense issues said that "apart from the useful and necessary compatibility with the requirements of the revision of the GMDSS, the defined application specific messaging, and other applications to facilitate entering and leaving port, and the associated shore side support, what is not yet seen, (which incidentally they felt was the biggest single demand from the shipping community), is how the e-Navigation and maritime cloud concept will help significantly reduce paperwork and administration."

Last, a manufacturing company who stated that they "think the opportunities for themselves are very limited...unless something like the VHF Data Exchange System (VDES) is included as part of e-Navigation. With VDES surfacing as a future maritime communication solution (maybe with VHF coastal stations capable of receiving VDES data), this company feel that there is a probable business case here, in combination with AIS."

If e-Navigation can be shown to save money for stakeholders, such as shipowners and port authorities, then they will drive the sales and availability of the equipment. It is important in this case that e-Navigation is well promoted by all concerned.

1.3.5 Ownership of e-Navigation

The e-Navigation concept involves a broad range of navigation systems and services and will require greater emphasis on the accuracy, quality, reliability, integrity, and timeliness of existing information services that are provided. Demands for harmonized, real-time information that meets international standards, such as water depth and level, buoy status, ice conditions, and weather forecasts, will be expected. Stakeholders need to ensure that their internal operations could meet these demands in a continuous, reliable, accessible, and fail-safe environment. Taking this technological turn could become an expensive and time-consuming process. For example, in migration from traditional systems to e-Navigation, there could be pressure to continue supporting both traditional and newer technologies. A pragmatic assessment of what is expected for products and time frame is on the agenda; however, user interaction should be focused on the need for e-Navigation products and services and should serve as the driver for any future investment. Properly implemented, e-Navigation should also enhance the best practices of traditional navigation by using human and machine decision-making processes to complement one another. It is also clear that also aids to navigation (AtoN) will have a part to play in further e-Navigation development.

Service providers and manufactures must be in a position to adapt their businesses to the new reality of better informed mariners. A constant reassessment of the need for traditional aids to navigation, such as type of, number, and placement given the capabilities of e-Navigation, would be needed so that authorities and other providers could redirect its investment in areas of most demand. Authorities on national and regional levels, institutional research funding, and industry have mostly financed the costs associated with the development of e-Navigation.

The main driver behind the work is expected benefits and business opportunities of e-Navigation. Each of the e-Navigation solutions has undergone risk- and cost-benefit analyses based on agreed IMO methodology. For shipborne and shore-based users, e-Navigation will simplify daily work and training, improve the human-machine interface, usability, familiarity, and navigational safety, and improve time management and efficiency on board. This will in turn provide easier access to information, thereby improving the response time and problem-solving abilities of bridge personnel. e-Navigation will also enhance the quality, accuracy, and reliability of information, thereby improving situational awareness and navigational safety, and improve familiarity with systems through standardization. For VTS and port services, e-Navigation will improve service and safety in regulated areas by providing easy access to available services and warnings and reduce the risk of accidents. This will also increase the effectiveness of traffic management and port operations.

During the development of e-Navigation, many have raised the question of whether the IMO should continue to take ownership and coordination of the development of e-Navigation.

Ownership of e-Navigation includes leading the development and maintenance of the vision, the concept and definition of services, and the concept of operations. A part of the ownership is also identification of responsibilities for the design, implementation, and operation of e-Navigation, and acknowledging the rights, obligations, and intentions of flag states, port states, and the various authorities within those states. In 2016, the IALA submitted a proposal to the IMO's Maritime Safety Committee (MSC), saying that it would be willing to contribute to the coordination of the work related to the development of e-Navigation MSPs, which is shore-based digital services provided to ships. However, the IMO agreed to keep the coordination of this subject under the scope of the organization, through the Navigation, Communications & Search and Rescue (NCSR) Sub-Committee.

The need for IMO control and leadership has been highlighted several times. The means of doing this is by incorporating e-Navigation into the IMO's future work plan. The International Chamber of Shipping (ICS) has stated that without continued IMO leadership of e-Navigation, there is a very real danger that a global approach will fail and that the project's momentum will be dispersed into a variety of uncoordinated regional initiatives. This has the potential risk that e-Navigation could be implemented differently, for example, in the Baltic region, the Singapore and Malacca Straits, and in North America. This could result in ships being required to have different systems for different regional e-Navigation services.

In February 2005, Mr. Brian Wadsworth, director of the Logistics and Maritime Transport of the U.K. Department for Transport, presented an orientation paper that introduced e-Navigation.

"Making a commitment to agree a vision for and create a migration plan to e-Navigation would be a strong signal of the commitment of governments and regulatory authorities to a positive agenda for the growth and development of the global shipping industry, whilst achieving key safety and environmental goals" [6].

Mr. Wadsworth concluded that a single international institution must take "lead ownership" of the vision for e-Navigation. This institution must have the ability to integrate and manage the essential contributions of key stakeholders in such a way as to promote constructive development and standardization. The logical choice of lead institution is the IMO.

The orientation paper from the United Kingdom, and further development of the ideas, resulted in a proposal on development of an e-Navigation strategy submitted to MSC from Japan, Marshall Islands, the Netherlands, Norway, Singapore, United Kingdom, and the United States in 2006 [7].

The authors of the submitted paper believed that it was now appropriate that the IMO developed a broad strategic vision for incorporating the use of new technologies in a structured way, ensuring that their use is compliant with the various electronic navigational and communication technologies and services that are already available. The aim was to develop an overarching accurate, secure, and cost-effective system with the potential to provide global coverage for vessels of all sizes.

It also highlighted the fact that there were already a great many electronic navigational and communication technologies and services available or in development. However, if such technological advancement remains uncoordinated, there is a risk that future development of the global shipping industry would be hampered through lack of standardization on board and on land, incompatibility between vessels, and an increased and unnecessary level of complexity.

"Our agenda is making navigation safer and more efficient for mariners and shore-based personnel, and I believe e-Navigation will have an important role to play in this. We live in a digital age and in order for us to evolve and grow we have to embrace and adapt to the digital evolution. This is the central core of e-Navigation which has already begun to make its mark on the maritime world, reshaping the way we navigate, communicate and provide services. As this development continues, we must not forget that e-Navigation, in the end, is about making it better for mariners to navigate safely across the seas, and to improve the efficiency of maritime transport and shore based services," concluded Director General Kirsti Slotsvik of the Norwegian Coastal Administration, in conversation with the author 10 years after the submission of the 2006 paper to IMO.

1.3.6 Concerns about e-Navigation

There is no doubt that some delegations at the IMO were skeptical of how mariners could benefit from the continuous technological development that has moved rapidly in our digital world. One example of such a discussion took place in the IMO Correspondence Group (CG) on e-Navigation when providing input to the SIP. It started with an article distributed by a member representing an international organization, titled "E-Navigation CG: Something to Think About." The article had obvious relevance to the e-Navigation movement and the work of the CG and was published in 2011 by The Associated Press. The heading was "Are Airline Pilots Forgetting How to Fly?" [8].

"As planes become over more reliant on automation to navigate crowded skies, safety officials worry there will become more deadly accidents traced to pilots who have lost their hands-on instincts in the air. Hundreds of people have died over the past five years in 'loss of control' accidents in which planes stalled during flight or got into unusual positions that pilots could not correct. In some cases, pilots made the wrong split-second decisions, with catastrophic results—for example, steering the plane's nose skyward into a stall instead of down to regain stable flight," the article said in its introduction.

It continues on: "....Pilots use automated systems to fly airliners for all but about three minutes of a flight: the takeoff and landing. Most of the time pilots are programming navigation directions into computers rather than using their hands on controls to fly the plane."

In preparation of the e-Navigation CG for one of the IMO sessions by the author, a representative of one of the major flag states at IMO responded: "This is a timely reminder of over reliance on automated systems. ....In shipping we have a similar situation in the engine room. Engineers have fewer opportunities to work on the ship's machinery due to increased reliability, more automation and better planned maintenance by shore personnel," and "The navigator must be given a much closer connection with the actions on the ship than reading information from a screen. The need to look out the window to assess the conditions around the ship must remain an essential part of bridge watchkeeping."

Another member of the CG stated that "it may be an idea to think about developing an (IMO) Assembly resolution raising this issue and urging the Membership to exercise caution not to allow technology to drive and or be ahead of ourselves when developing future solutions."

Fortunately, the proposed retroactive Assembly resolution was not developed. Being responsible for establishing mandatory standards for enhancing the safety of life, maritime security and protection of the marine environment, as well as having a global remit, IMO is the only organization that is capable of meeting the overall governance and ownership of e-Navigation. The responsibility that comes with IMO ownership and control of the concept has been accepted by IMO member states.

Because of the safety and environmental issues surrounding ship navigation and the global nature of the IMO, it is important that technological advances are coordinated in terms of systems and equipment interoperability, and wherever possible, standardization of procedures. e-Navigation should not alter the responsibilities of the navigator in terms of safe navigation of the ship.

The e-Navigation concept is not limited to the navigator and intership communications. Ship and shore voice and data communications will mean that e-Navigation will affect the entire maritime industry, including port and national authorities as well as ship agents and equipment manufacturers. It will also be a scalable system across vessel sizes and types.

1.4 Aims and Objectives of e-Navigation

Moving a ship from port to port is not an exclusive navigational process. There are other processes involved, such as ship reporting, practical and economic cargo handling, cargo control, legal requirements, and security.

It is well recognized that IMO's e-Navigation concept—particularly the provision of reliable and timely data and information dissemination along with enhanced interaction between ship and shore—could contribute to enhancing the safety of navigation, make improvements in the efficiency of shipping, provide better access to sea areas and ports, and allow further development of a worldwide, sustainable maritime transportation system.

1.4.1 Safety Including Reducing Accidents

e-Navigation is expected to improve safety through promotion of standards in safe navigation supported by improved decision support. This will enable mariners and competent authorities ashore to select unambiguous information pertinent to the prevailing circumstances, in addition to reducing human error through the provision of automatic indicators, warnings and fail-safe methods. Through e-Navigation, standardized equipment with an S-mode option has been introduced. S-mode is a proposed functionality for shipborne navigation displays using a standard, default presentation, menu system, and interface, but without restricting the ability of manufactures to innovate. The enhancement of navigation system resilience will result in improved reliability and integrity and better integration of ship- and shore-based systems, and will in turn lead to better utilization of all human resources that will also improve safety and reduce accidents. According to IMO's e-Navigation strategy plan, a rising trend of marine accidents, both in terms of numbers and costs, is mainly associated with collisions and groundings.

Captain Jorge Pecci, executive vice president and head of Global Marine with American International Group, Inc. (AIG Americas), said that despite progress, "the human element" remains the most important factor in maritime accidents, and navigational errors continue to account for large losses. Figure 1.3, presented by AIG, shows serious losses from 1998 to 2012 by cause, based on the frequency in percent of all serious and losses for that period. The frequency of collisions and groundings are marked with ellipses [9].

"In order to reduce navigational error, we must ensure that new technologies can aid, rather than hinder the mariner. Navigation systems, vessel traffic information and communication infrastructure must work as efficiently as a Swiss watch. Such harmonized information should enhance situational awareness and decision support for mariners and consequently, enhance navigational safety and efficiency," Captain Pecci ascertained [9].

According to the IMO strategy plan on e-Navigation, around 60% of collisions and groundings are caused by direct human error, and there are numerous examples of collisions and groundings that might have been avoided had there been suitable input to the navigation decision-making process. "Despite advances in bridge resource management training, it seems that the majority of watch-keeping officers make critical decisions for navigation and

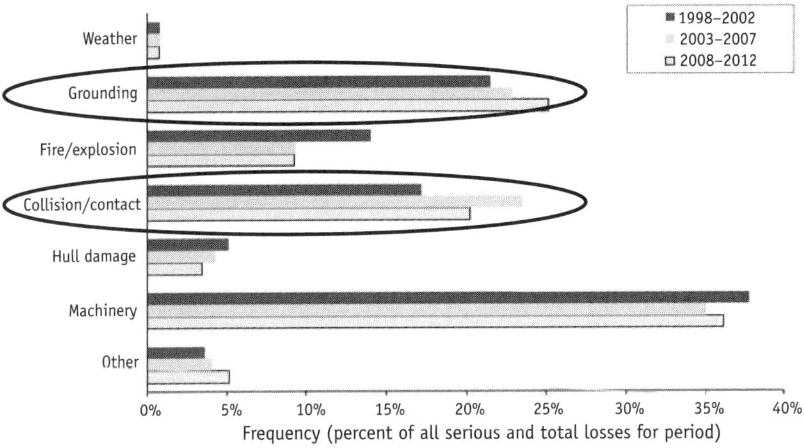

Source: LMIU, total losses as reported by Lloyds List

Figure 1.3 Serious losses, by cause, of all types of vessels above 500 GT, 1998–2012.

collision avoidance in isolation, due to a general reduction in human resources" [10].

1.4.2 Efficiency and Reduced Costs

More energy, food, and leisure resources will be required to satisfy the growing population of the world. Most of these demands will be met in some form using maritime transport, especially container ships. e-Navigation is expected to equip shipboard users, and those ashore responsible for the safety of shipping, with effective, user-friendly, proven tools that are optimized for effective decision-making in order to make marine navigation and communications more reliable, resilient, and user-friendly.

Increased efficiencies and reduced costs (aided by dynamic route planning, sea traffic coordination, reduction of steaming distances, coordinated arrival times, and information transfer to logistics stakeholders) are some examples where e-Navigation will contribute to the global economy.

1.4.3 Use of e-Navigation in Security

Another benefit of e-Navigation is augmented security, which takes place by enabling a more silent operation mode for shore-based stakeholders for domain surveillance and monitoring. The integration of systems that are already in place, precipitating efficient and coherent use of new equipment that meets all user requirements, could result in higher efficiency and reduced costs.

The importance of global maritime transport is well recognized, but there is also a growing sensitivity to, and awareness in many countries of, the importance of the marine environment and the need to actively protect coastal waters. There is an increasing intolerance of adverse impacts on the marine environment caused by shipping, and in particular accidents that result in marine pollution. Thus, there is a need to adopt appropriate strategies for monitoring and surveillance of shipping and its interface to a wide range of data information, along with reporting systems that support marine navigation. Many of these applications have direct relevance to the IMO e-Navigation strategy plan.

An example of integrated systems comparable to those in e-Navigation is the European Agency Frontex, which "promote[s], coordinate[s] and develop[s] European border management in line with the EU fundamental rights charter applying the concept of Integrated Border Management, similar to ships monitoring provided by many coastal states" [11].

"Frontex helps border authorities from different EU countries to work together. Frontex's full title is the European Agency for the Management of Operational Cooperation at the External Borders of the Member States of the European Union. The agency was set up in 2004 to reinforce and streamline cooperation between national border authorities" [12]. In pursuit of this goal, Frontex has several operational areas.

Information systems and the information sharing environment is one of the areas that the e-Navigation philosophy of integration will support. Information on emerging risks and the current state of affairs at external borders forms the basis of risk analysis and situational awareness for border control authorities in the EU.

The services include automated vessel tracking and detection capabilities, software functionalities allowing complex calculations for detecting anomalies and predicting vessel positions, as well as precise weather and oceanographic forecasts. They are closely related to those in e-Navigation, such as ship reporting, and some maritime services provided from shore, such as from VTS centers.

Security also includes other operations such as piracy. The shipping world is still under threat from pirates in West and East Africa, with new hotspots emerging in South East Asia. The knowledge of ships' locations in a secure manner with integrity, which is an important principle of e-Navigation, helps prevent piracy.

According to the Global Maritime Security Market 2015–2019 report, "maritime security operations cover missions ranging from preventing shipping of illegal goods to countering piracy. They also involve the prevention of embargo situations and human trafficking. Maritime security agencies protect the nation from illegal activities and threats, such as smuggling, drug trafficking, piracy at sea, human trafficking, and maritime terrorism. These agencies provide security in ports and for marine traffic. Maritime terrorist activities, such as the Mumbai attack in 2008 and the rise in piracy in the Gulf of Aden, has raised concerns about the need to secure ports, and modernize weaponry and equipment utilized by maritime security agencies" [13].

Modern proven tools and integrated systems for increased situational awareness are crucial in this respect. These views are similar to e-Navigation principles.

Maritime manufacturing companies already produce displays showing different integrated data including equipment and electronic positioning fixing systems (EPFS). More companies are working on other benefits, such as under-keel clearance and collision avoidance. These integrated displays help to achieve optimum safety and efficiency, but also have economic benefits such as reduction in workload, best passage provision that reduces risk of groundings, hazard avoidance, lower insurance premiums, as well as environmental benefits such as reduction in fuel consumption.

1.4.4 Use of e-Navigation and Cybersecurity

There are pluses and minuses relating to digitalization and the use of cyberspace. It increases the number of potential threats, but also provides the potential to allow for greater automated vigilance on existing threats.

Many stakeholders are aware of the possible threat to e-Navigation systems from cyberattacks, and have recognized the need for standards and guidelines for ensuring cybersecurity on ship systems. Critical infrastructure—the physical and virtual assets, systems, and networks vital to national and economic security, health, and safety—are vulnerable to cyberattacks by foreign governments, criminal entities, and lone actors. Cyberattacks are often identified as the greatest threat to national security. Due to the increasingly sophisticated, frequent, and disruptive nature of cyberattacks, such an attack on critical infrastructure could be significantly disruptive or even potentially devastating.

The IMO has developed high-level and nonmandatory guidelines on cyber risk management with a focus on operational aspects, *Interim Guidelines on Maritime Cyber Risk Management*. These guidelines support shipowners in establishing procedures, plans, and instructions, including checklists as appropriate, for relevant key shipboard operations. Such instructions, plans, and procedures should be seen as complementary to existing security and safety risk management requirements, contained in the International Safety Management (ISM) and The International Ship and Port Facility Security (ISPS) Codes [14].

Cyberthreats are a growing menace, spreading to all industry sectors that rely on information and communications technology (ICT) systems, including the maritime sector. Deliberate disruptions of critical automation systems and cyberattacks in general have a significant impact on critical infrastructures. The Australian Maritime Safety Authority (AMSA) has reported that threats are more frequent than we think. In a typical two-hour period there are numerous attempts made to break into Australian sites, both governmental and private. Australia attracts a disproportionately large number of attacks, as a lot of its business is carried out online. Also, Australia protects substantial important intellectual property (e.g., WiFi).

Due to the complexity of ICT, it is a major challenge to ensure maritime cybersecurity. During the development of e-Navigation, several parties encouraged a common strategy and development of good practices for the use of ICT systems, in order to ensure security by design for all critical maritime ICT components on board and ashore.

The IMO's work on the implementation of e-Navigation systems and equipment includes addressing the need to ensure ICT robustness against cyberattacks. Undertaking awareness-raising campaigns that target the maritime sector and cybersecurity training for shipping companies, port authorities, navigators and vessel traffic service operators (VTSOs) were highly recommended. In addition, as current maritime regulations and policies generally consider only physical aspects of security and safety, policy makers should also include due consideration of addressing cyber security. Marine electronic manufacturers were also encouraged to address measures to allow recovery once an attack has occurred and to develop strategies of detecting ongoing attacks. ICT systems should always be secure to ensure user confidentiality and data integrity based on a safe communication network.

In this respect, work led by Australia, Japan, and the ROK, in cooperation with international organizations such as the Nautical Institute (NI) and the manufacturers' CIRM , provided a common approach "to ensure that software trustworthiness and user needs are met, through the application of software quality assurance (SQA) and human-centered design (HCD) in the development of e-Navigation systems and equipment." In 2015, MSC 95 approved the results of this work, titled *The SQA and HCD Guideline for e-Navigation* [15].

The increasing use of the Internet of Things (IoT) and its application to onboard systems means that security is not limited to the maritime sector but is a universal problem for all vital digital services. Everyone in the chain has a responsibility for appropriate security, such as manufacturers, shipowners, service providers, and national and local authorities.

According to the strategy plan, the IMO agreed that e-Navigation systems should be resilient and take into account issues of data validity, plausibility, and integrity in order to be robust, reliable, and dependable. Requirements for redundancy, particularly in relation to position-fixing systems, should be considered. Position-fixing systems will need to be provided to meet user needs in terms of accuracy, integrity, reliability, and system redundancy, in accordance with traffic risks and volume [10].

The Global Positioning System (GPS) has become commonplace, supporting critical infrastructure depended on by many across all industries, including the shipping industry. It was originally developed to provide precision navigation and timing to the United States military. Today, however, society has rapidly developed a strong reliance on GPS and other Global Navigation Satellite Systems (GNSS) information in many facets of life, including transportation, safe navigation, communications, and financial transactions. GNSS has become an indispensable source of information with significant economic benefits, making it increasingly important that GNSS data is widely available, secure, and reliable.

In 2014, Jeffrey Coffed with the U.S. company Exelis wrote in the article "The Threat of GPS Jamming" that "GPS is susceptible to service outages due to intentional and unintentional jamming. To maintain critical infrastructure, proactive measures must be taken to eliminate GPS disruptions. Addressing the security of GPS signals and preventing denial of service should be a priority. Detection and tracking is the best and first step to take as the availability and usage of low-cost GPS jamming devices has resulted in an increased threat to GPS signal disruption." This also applies to other current and future GNSS such as the Russian GLONASS, the European Galileo system, and the Chinese BeiDo system [16].

1.4.5 Protection of the Environment

e-Navigation focuses on better protection of the environment by reducing emissions while using optimum routes and speeds. e-Navigation encourages the exchange of marine information onboard and ashore by electronic means during the entire voyage of a vessel. Several projects have focused on this issue, including the SESAME Straits project: "Secure, Efficient, and Safe maritime traffic Management in the Straits of Malacca and Singapore, which aim[ed] to utilize the IMO e-Navigation concept in order to reduce the problem of traffic hot-spots" [17].

The objective of SESAME Straits was to optimize sea traffic flow and provide vessels with advice on the best speed and route. This enables the vessel to arrive at a port at the optimal time, or create an optimal transit passage. In this way it is possible to avoid traffic congestion, reduce fuel consumption, reduce emissions, and increase overall safety and efficiency. "This will improve the safety of vessel traffic and enable 'just-in-time' arrival of vessels, improving the efficiency of existing infrastructure, and reducing the environmental footprint" [17].

This project was established in parallel with the development of the e-Navigation SIP and "serves as a highly important test bed for the IMO's e-Navigation concept. It intends to test various aspects of e-Navigation and to facilitate the implementation of e-Navigation solutions in the maritime domain" [17].

"For this reason, the project [received] guidance from the Straits e-Navigation Alliance (SENA) High Level Advisory Board (HLAB), which consist[ed] of members from IMO, IHO [International Hydrographic Organization], IALA [The International Association of Marine Aids to Navigation and Lighthouse Authorities], ICS [The International Chamber of Shipping], BIMCO [The Baltic and International Maritime Council], as well as the maritime administrations of Indonesia, Malaysia, Norway, and Singapore" [17]. In this respect, the MSC has also approved guidelines on harmonization of test beds reporting. Test beds are important to investigate whether the operational and technological concepts work and function together as provided. A test bed is a platform for trialing development projects. Such e-Navigation test beds allow for early detection of new system functionality, operation usability, areas of enhancements, and identification of weaknesses. More about test beds are described in Chapter 4.

The years after the IMO had approved the e-Navigation SIP, Mr. Fred Pearce, award-winning science writer, presented an article in the UK's MailOnline titled "How 16 Ships Create as Much Pollution as All the Cars in the World."

"As ships get bigger, pollution is getting worse. The most staggering statistic of all is that just 16 of the world's largest ships can produce as much lung-clogging sulphur pollution as all the World's cars. Because of their colossal engines, each as heavy as a small ship, these super-vessels use as much fuel as small power stations. But, unlike power stations or cars, they can burn the cheapest, filthiest, high-sulphur fuel: the thick residues left behind in refineries after the lighter liquids have been taken. The stuff nobody on land is allowed to use," Pearce claimed [18].

There are now an estimated 100,000 ships on the seas, and the fleet is growing fast as goods are ferried in vast quantities from Asian industrial powerhouses to consumers in Europe and North

America, but according to Mr. Pearce, burning low-sulfur fuel will not cut carbon emissions from ships.

"But there are other ways. More efficient engines could reduce emissions by 30 per cent," according to British marine consultant Robin Meech. "Cutting speed could reduce emissions by as much again. And there are even wackier ways, such as putting up giant kites to harness the wind as in the days of sailing ships," he proposed [19].

Improved berth-to-berth navigation, as described in the original definition of e-Navigation, will cut emissions especially in relation to just-in-time arrival. Enhancing the protection of the environment and taking into account the need for global application is one of the drivers behind the need for e-Navigation. As the improvement of navigation safety leads to better protection of the environment, it will reduce the risk of collisions and groundings, along with the associated oil spillage and pollution, as well as enhance the ability and capacity in responding and handling emergencies such as oil spills.

According to the World Wildlife Fund (WWF), substandard ships and poor shipping practices are leading to massive marine pollution and damage. "The threats posed by shipping are not spread evenly across the oceans, but rather concentrated in busy shipping lanes and ports. As shipping lanes become more and more congested, the level of pollution caused by shipping increases - as does the chance of oil spills and accidents [...] International law limits the ability of coastal nations to impose and enforce their own environmental and navigation regulations on foreign ships passing through their waters. Instead, countries must use international conventions established through the International Maritime Organization (IMO) and the UN Convention on the Law of the Sea (UNCLOS)," WWF recommends [20].

Various stakeholders called for the environmental effects of shipping to be put in line with the text agreed on at the COP21 climate talks in Paris in 2015. The stakeholders claimed the need to act decisively in order to further regulate CO2 emissions from ships on a global level. They stated that it would provide support for their ability to move forward and give new impetus to ongoing discussions. Shipping, together with aviation, had been included in earlier versions of the draft agreement on climate change. The latest version in 2015, however, makes no reference to these two industries.

The IMO has already made strides when it comes to the shipping industry's carbon footprint. In 2011, the IMO adopted universal measures that made shipping the first industrial sector to have a binding global regime in place to reduce its CO2 emissions. "Entered into force in 2013, the measures require a gradual improvement of energy efficiency for newly built ships through intermediary targets, culminating in the requirement that all ships constructed after 2025 be 30 per cent more efficient that those built in the 2000s." The implementation of e-Navigation will support this initiative [21].

According to the European Commission, emissions of air pollutants like sulfur dioxide can travel long distances, and in recent years emissions from maritime transport have increasingly affected air quality in the EU. As of January 1, 2015, EU Member States have to ensure that ships in the Baltic, the North Sea, and the English Channel are using fuels with a sulfur content of no more than 0.10%. Higher sulpfur contents are still possible, but only if the appropriate exhaust cleaning systems are in place. Sulfur dioxide emissions cause acid rain and generate fine dust. This dust is dangerous for human health, causing respiratory and cardiovascular diseases and reducing life expectancy in the EU by up to two years [22].

1.4.6 Global and Technical Standardization

The provision and developments of e-Navigation should consider relevant international conventions, regulations and guidelines, national legislation, and standards. The development and implementation of e-Navigation should build upon the work of the IMO, including, but not limited, to the requirements prescribed in the International Convention for the Safety of Life at Sea (SOLAS), the International Convention for the Prevention of Pollution from Ships (MARPOL) and the International Convention on Standards of Training, Certification and Watchkeeping for Seafarers (STCW).

IMO's strategy plan for e-Navigation emphasizes that if current technological advances continue without proper coordination, there is a risk that the future development of marine navigation systems will be hampered through a lack of standardization. This could result in incompatibility between vessels and an increased and unnecessary level of complexity on board and ashore.

e-Navigation is intended to meet present and future user needs through harmonization of marine navigation systems and supporting shore services. IMOs Strategic Plan for the Organization for the period of 2008 to 2013 recognized that technical developments have created new opportunities, but may also open a window to new threats such as cyberattacks. New opportunities therefore existed to further develop various IMO initiatives, from safety and security to environmental protection.

The challenge for IMO was to ensure that the technical developments adopted are conducive to enhancing maritime safety, security, and protection of the environment, and take into account the need for their global application. Furthermore, the challenge is also to ensure the proper application of information technology within the IMO and to provide enhanced access to that information for the shipping industry and others. This should in turn ensure that new equipment for use on board ships is designed and manufactured with the needs, skills, and abilities of all users in mind. Having said that, improved safety, through promotion of standards in safe navigation, should lead to higher efficiency and reduced costs enabled by global standardization. The IMO was also of the opinion that the availability of ENCs worldwide was most important and requested IHO and Member Governments to continue their efforts in increasing the coverage.

e-Navigation will likely benefit from increased functionality of the future IHO S-100 Universal Hydrographic Data Model. Furthermore, IMO Member States have a responsibility for the safety of all classes of vessels. This may include the scalability of e-Navigation for all potential users. Extension of the concept to non-SOLAS vessels' scalability should be seen as an important task to be addressed in the first instance through consultation on user requirements.

e-Navigation should be forward- and backward-compatible, and support integration with equipment and systems made mandatory under international and national carriage requirements and performance standards. The highest level of interoperability between e-Navigation and external systems should be sought where practicable.

A core objective of e-Navigation is to facilitate global coverage, consistent standards and arrangements, and mutual compatibility and operability of the equipment, systems, symbology, and operational procedures. This is to avoid potential conflicts between users and to support scalability for all ships, including non-SOLAS ships, facilitating the use by all potential maritime users.

1.4.7 Communications

Any communications systems should be able to provide reliable data communications between ship and shore and vice versa.

The existing available communications can be broadly divided into the following categories:

1. Used for distress and safety related communications, such as for the promulgation of Maritime Safety Information (MSI), as is currently mandated by Global Maritime Distress Safety System (GMDSS), and Automatic Identification System (AIS).
2. Commercially available systems, such as various satellite solutions for example, Inmarsat, Iridium, and very small aperture terminal (VSAT), as well as terrestrial telephone and data networks such as Global System for Mobile Communications–3G/4G (GSM) [23].

Existing and future communication links could be integrated via a maritime intranet, although each technical service will be limited by the capabilities of the utilized communications link. This infrastructure will primarily be based on Internet Protocol (IP) communications links but will enable the utilization of cost-free communication links for safety and mandatory reporting where appropriate, enabling a seamless integration and transition between available communications technologies.

Relevant requirements for commercial communication links for e-Navigation should have certain availability and latency criteria for the defined service area, and should provide a two-way data communication channel, enabling acknowledgment of information delivery.

This could enable automatic quality assurance of

- Service efficiency;

- Availability and coverage of the communication service;
- Quality of the shipborne communication installation and capability.

It is envisaged that the majority of communication of shore data would be needed as a vessel approaches the coast, and therefore it is likely that more bandwidth/speed may be needed in these areas.

There is already a limited but broadband network linking the long-range identification and tracking (LRIT) data centers in many countries, which might be used for communication of ships' data between countries. The LRIT system currently provides tracking information of ships to IMO Member States.

Future systems could include very high frequency (VHF) data or VDES and navigational data (NAVDAT) and be developed for Internet-based solutions, such as a maritime cloud, facilitating system-wide information management solutions. The Danish developed Maritime Cloud is a digital information technology (IT) framework consisting of standards, infrastructure, and governance that facilitates secure interoperable information exchange between stakeholders in the maritime community using the principles of service-oriented architectures (SOA). The core of the Maritime Cloud consists of three key infrastructural components providing central framework services:

1. The *Maritime Service Portfolio Registry* is an encounter point for those that consume, provide, or specify services in the maritime domain. It enables service standardization and automatic service provision.
2. The *Maritime Identity Registry* provides all maritime stakeholders with a basic maritime identity and basic methods for authentication, integrity, and confidentiality in information transfer through the use of digital certificates in a public key infrastructure (PKI).
3. The *Maritime Messaging Service* is a geo-aware messaging service that takes into account the needs of ships in terms of achieving interoperability across varying data links with varying availability, technical characteristics, and limited bandwidth [24].

The concept of the Maritime Cloud represents an important future digital infrastructure for practical implementation of global e-Navigation. Its development, including funding, operational and legal issues, liability, quality, and accessibility of information and global functional operation, have to be studied before being accepted.

1.4.8 Training and Familiarization

The e-Navigation strategy identified training, competency, language skills, workload, and motivation as essential. Alert management, information overload, and ergonomics are primary concerns. These aspects of e-Navigation were also taken into account in accordance with IMO's Human Element work.

In the broadest sense, e-Navigation should support users to manage information to ensure good situational awareness and facilitate consistently good decision-making. Looking at the challenges in the industry, it was important to recognize opportunities offered by advancement in technology. It is also expected that ships will become increasingly bigger and more complex, and seafarers may be expected to have the qualifications and competency to meet these developments. Still, it should be clear that the major challenge to address is the human element.

The IMO Human Element Analyzing Process (HEAP) was essentially a checklist for issues to consider, in particular relating to human element, organizational, and training issues. The human element was also an important factor in the e-Navigation gap analysis, as this was divided into four elements: technical, operational, regulatory, and training. More about the IMO gap analysis is described in Chapter 3.

Seafarer qualifications should address the use of technology, the speed of its introduction, and the different needs for training. This was part of the more general development of e-Navigation that required not only documentation of skills, but also of how seafarer competence is to be maintained and updated. There may be a need for people to have different competences onboard and ashore, but their ability to operate interactively is important.

It is fundamental that a framework for harmonizing maritime information systems must be designed to facilitate the reduction of single-person errors/single-unit errors, particularly onboard. This requires the system to reduce some of the basic errors in perception, communication, and decision-making that can occur onboard as well as ashore. Key requirements in relation to human/machine interface included more effective means of analyzing and prioritizing information under both normal and abnormal conditions. A strong need for improved language skills, particularly competency in maritime English by all users, has been emphasized. Another important element was the need for a standardized shipborne navigation display to improve effective use and to enhance training.

There should also be equipment performance standardization, including a standard mode of operation for shipboard equipment (S-Mode), as previously mentioned, and the software installed in operating systems should follow a formal change control process to ensure that all elements of the e-Navigation system will operate efficiently. The development of S-Mode for bridge equipment might facilitate improved movement of crew between vessels.

The development of a standardized, module-based bridge layout adapted to the functions of the individual ship should be considered in order to facilitate the smooth familiarization of shipborne personnel when transferring from one ship to another. It may be further noted that potential stakeholders are engaged in an ongoing development of configurable, standardized data modules that could connect basic functions. In this context, training implications should also be considered.

In the future, it may be even more important to standardize the required levels of competence of maritime instructors. This may have an impact on the recruitment of adequate instructors.

e-Navigation is a continuous process of reviewing user needs, identifying cost-effective solutions, and is about improving what we have today and keeping up with new technology as it advances. Nevertheless, at the end of the day, e-Navigation should lead to safer, more efficient navigation and reduction of administrative burden.

References

[1] Lim, K., keynote speech, e-Navigation Underway 2016, transcript [online], available at http://www.imo.org/fr/MediaCentre/SecretaryGeneral/SpeechesByTheSecretaryGeneral/Pages/iala.aspx, accessed May 10. 2017.

[2] IMO document NAV 54/25 Annex 12.

[3] IMO MSC 94/21, paragraph 9.15.

[4] Information provided by Kwan An, former deputy director, Ministry of Oceans and Fisheries in the Republic of Korea, via personal communication, available at http://www.thedigitalship.com/news/electronics-navigation/item/2925-korea-to-invest-200m-in-e-Navigation#ixzz4gmGV0OHg.

[5] ACCSEAS Final Report, 13/05/2015 [online], available at http://www.accseas.eu/content/download/8190/71102/ACCSEAS%2520Final%2520Report%2520v1.pdf, accessed May 10, 2017.

[6] Wadsworth, B., "Marine eNavigation: An Orientation Paper," March 10, 2005, available at https://www.iho.int/mtg_docs/circular_letters/english/2005/Cl29e.pdf, accessed May 10, 2017.

[7] Maritime Safety Committee, IMO, "Development of an e-Navigation Strategy," Document MSC 81/23/10, December 19, 2005 [online], available at https://www.iho.int/mtg_docs/com_wg/HSSC/CHRIS18/CHRIS18-04.4C_e-Navigation_strategy.pdf, accessed May 10, 2017.

[8] Lowy, J., "Are Airline Pilots Forgetting How to Fly?" August 30, 2011 [online], available at http://www.nbcnews.com/id/44324527/ns/business/t/are-airline-pilots-forgetting-how-fly/, accessedMay10, 2017.

[9] Pecci, J., E-Navigation [presentation online], 2015, available at https://iumi.com/images/Berlin2015/3Pressies/1609_JorgePecci_e_Navigation.pdf, accessed May 10, 2017.

[10] IMO, Strategy for the Development and Implementation of E-Navigation, Annex 20, Document MSC 85/26/Add.1, available at http://www.imo.org/en/OurWork/Safety/Navigation/Documents/enavigation/MSC%2085%20-%20annex%2020%20-%20Strategy%20for%20the%20development%20and%20implementation%20of%20e-nav.pdf, accessed May 10, 2017.

[11] Frontex Missions and Tasks [online], available at http://frontex.europa.eu/about-frontex/mission-and-tasks/, accessed May 10, 2017.

[12] Definitions from FRONTEX [online], available at http://bev.berkeley.edu/refugees/definitions.html, accessed May 10, 2017.

[13] Sandler Research, Global Maritime Security Market 2015–2019, November 2015.

[14] IMO, International Safety Management (ISM) and The International Ship and Port Facility Security (ISPS) Codes, IMO documents MSC 96/4/1 and MSC 96 WP1/Add 1.

[15] IMO, The SQA and HCD Guideline for e-Navigation, document MSC.1Circ.1512, 2015.

[16] Coffed, J., 'The Threat of GPS Jamming," January 2014 [online], available at http://gpsworld.com/wp-content/uploads/2014/02/ThreatOfGPSJamming_FEB14.pdf, accessed May 10, 2017.

[17] Sesame Straits, A Brief Overview of the Sesame Straits [online], available at http://www.straits-stms.com/Project.html, accessed May 10, 2017.

[18] Pearce, F., "How 16 Ships Create as Much Pollution as All the Cars in the World," *DailyMailOnline,* November 21, 2009 [online], available at http://www.dailymail.co.uk/sciencetech/article-1229857/How-16-ships-create-pollution-cars-world.html#ixzz3t9fXrqAW, accessed May 10, 2017.

[19] Meech,R., cited in "As 50 Giant Oil Tankers Lurk Off the British Coast, We Reveal an Even More Scandalous Threat to the Global Environment," *The Mail on Sunday,* November 22, 2009.

[20] WWF, Marine Problems: Shipping [online], available at http://wwf.panda.org/about_our_earth/blue_planet/problems/shipping/, accessed May 10, 2017.

[21] *Maritime Executive* Staff, "Shipowners Push for IMO Control of CO2," *The Maritime Executive,* December 1, 2015 [online], available at http://www.maritime-executive.com/article/shipowners-push-for-imo-control-of-co2, accessed May 10, 2017.

[22] European Commission, Transport Emissions, June 27, 2016, available at http://ec.europa.eu/environment/air/transport/ships.htm, accessed May 10, 2017.

[23] IMO document NCSR 1/28, Annex 7.

[24] ACCSEAS, Service Description: Maritime Cloud, available at http://www.accseas.eu/content/download/8217/74329/Service%2520Description%2520-%2520Maritime%2520Cloud%2520v1.pdf, accessed May 10, 2017.

2

Maritime Navigation: Current Equipment and Practices

2.1 Navigational Equipment, Systems, Displays, and Bridge Systems

The maritime sector is a key link in the global transport chain. In addition to the important navigational process, moving a ship from port to port also involves other processes such as practical and economic cargo handling as well as cargo control, legal requirements, and security. Yet, these various processes are relatively expensive and time-consuming for the shipping industry. The future will demand electronic integration of these processes.

Sometimes maritime accidents occur. The results of these accidents include potential loss of life and the immediate environmental effects of such accidents, loss of cargo, and loss or damage to ships. Moreover, the effects can present severe long-term problems for marine flora and fauna. However, most mariners hope that the future will see innovative designs, technologies, and working practices for safer and more efficient operations. This will reduce the environmental risks and potential loss of life, thus promoting more environmentally friendly ship operations [1].

Recalling the expectations of e-Navigation, which focuses on user needs, three areas are defined where technological developments and consequently changes are expected: on board, ashore, and the communications supporting e-Navigation in general. Looking at the situation today, however, there are several areas that cause confusion for the mariner. Let's look at some of them from the user's point of view.

The navigational equipment carried on board a ship is laid down by IMO SOLAS Chapter V, which was revised thoroughly in 2000 to take into account the developments in electronic position-fixing aids such as GNSS, typically GPS. Before that, requirements included a radio direction finder (RDF) as a radio navigational aid, but many ships carried terrestrial radio navigational aids, such as Loran and Decca, which were used where the service was available. And even before that, ocean navigation relied on celestial observations and dead reckoning, tried and tested methods used for centuries.

The basic navigational equipment required today depends on the size of the ship, and the rules include provisions for backup systems and alternates. Regulation 19 of SOLAS Chapter V states that all ships (on international voyages) must carry a magnetic compass and charts suitable for the intended voyage, which can be an ECDIS, as well as a navigation system receiver such as a GPS.

Over 300 GT, all ships must also carry an echo sounder to measure the depth of water and a 9-GHz radar to detect and measure the range and bearing of objects on the surface, as well as an electronic plotting aid to plot the range and bearing of targets to determine collision risk. They are also required to carry a speed- and distance-measuring device to measure the speed and distance through the water, which is known as a log. These devices need a transmitting heading device to work correctly. They are also required to carry an AIS.

Over 500 GT, ships are also required to carry a gyro compass with a repeater system for the transmission of heading information to other devices, as well as an automatic tracking system to plot the range and bearing of targets to determine collision risk. They are also required to carry an ECDIS by 2018.

Over 3,000 GT, ships are also required to carry a 3-GHz radar with an automatic tracking system, as well as the 9-GHz radar already required.

Over 10,000 GT, the automatic tracking aid must be able to track at least 20 targets and the ship must also be fitted with a heading or track control system to keep to a straight track or bearing (automatic pilot).

Over 50,000 GT, ships are also required to carry a rate-of-turn indicator as well as a speed- and distance-measuring device able to measure speed and distance in a forward and athwart ship's direction [2].

Other equipment that has an input to the safe navigation of a ship, such as NAVTEX and other means to receive MSI, is currently a carriage requirement as part of the GMDSS in SOLAS Chapter IV.

Other rules require ships to carry VDRs, but these are not considered here as they are not navigational aids.

SOLAS Chapter V does allow for integrated navigation systems and integrated bridge systems, but does not make them mandatory. This means that many ships carry individual pieces of equipment, but they are not integrated, meaning the navigators have to physically and mentally integrate the information they can see from the various devices to be sure that they have proper situational awareness as well as being aware of what they can see through the window. e-Navigation, by integrating and displaying the information in a better way, aims to improve this situational awareness and reduce the navigator's workload.

Today, with many different manufacturers of navigational equipment, menus and user design, including display and controls, differ from one product to another. A challenge faced by all mariners is to become quickly familiar with the wide range of systems and models of navigational equipment supplied by different manufacturers. Because of the increasing complexity and functionality of navigation equipment, a need has arisen for navigation systems to have more standardized functionality and layout to enable better operation to support good decision-making.

According to the American Pilots' Association (2016), a portable pilot unit (PPU) can be described as "a portable, computer-based system that pilots bring aboard a vessel to use as a decision-support tool for navigating in confined waters [...] Pilotage is, and always has been, a local matter. What works well for one district may be impractical or cumbersome for a geographically different area. And, in some cases, use of a PPU may not be appropriate" [3].

To take an individual view, in a personal interview, Captain Simon Pelletier, president of the International Maritime Pilots' Association and president of the Canadian Marine Pilots' Association, says that "every day marine pilots around the world are responsible for safely guiding vessels of all sizes and types through the world's most treacherous waterways. Like mariners in general, pilots are practical persons who, for centuries, have been making the most of every technological innovation that demonstrates it can make navigation safer and more efficient."

"Pilots have been both early developers and adopters of innovative e-Navigation [type] applications, such as PPUs, that have added a great deal of value to sensitive pilotage operations. In all cases, the focus of pilots has been on making sure that e-Navigation applications squarely put the emphasis on pragmatic deliverables, truly designed to meet the actual needs of mariners. Pilots have also found that keeping expectations realistic is essential to success: enhanced use of electronic navigational tools can deliver significant benefits, but it is important not to overstate benefits or underestimate the time required to fully implement new approaches. Just as important, success is achieved only through the concerted, collaborative efforts of all marine stakeholders, focusing on one objective at a time—and doing it well. The development of e-Navigation is a collective task and, when stakeholders feel their interests have been duly taken into account, they are more likely to support any given initiative," argues Captain Pelletier.

An interesting aspect in this respect is that a PPU itself represents an integrated system in line with the e-Navigation philosophy; however, the information provided only for the pilot on a PPU is not shared to or displayed on the ships' own information system on the bridge or any systems ashore, such as VTSs and ports. Some mariners have, however, raised questions whether PPU information should be more harmonized and shared between relevant equipment onboard and ashore and vice versa.

Many ships are now navigating oceans and waterways without paper charts. Figures presented by the United Kingdom

Hydrographics Office (UKHO) in 2015 showed that "of an estimated 41,500 internationally trading ships around the world, 24,300 or 58% are using an ENC (electronic navigational chart) service on ECDIS as a result of the SOLAS-mandated carriage of ECDIS, which is being introduced on a rolling timetable for different ship types and sizes [towards 2018]" [4].

Captain Harry Gale, FNI, technical manager at the NI feels that, "The ECDIS concept is a total change from using paper charts and the transition from paper charts to electronic charts will pose a challenge for the industry, particular for those who have no current experience of electronic charts" [5].

The NI's research has shown that those using electronic charts consider the positive aspects of ECDIS to be

- Easier voyage planning;
- Simpler chart correction;
- Continuous monitoring of depth safety contours and soundings;
- Readily available information when approaching busy ports or navigational areas.

However, they also identified areas mariners should be aware of when navigating using ECDIS [5], such as

- Too much information on the screen being distracting;
- Submenus being very complex;
- The size of charts displayed on the screen monitor being very much reduced compared with paper charts;
- Some symbols being open to misinterpretation due to unfamiliarity;
- Automatic position plotting leading to complacency.

"Electronic charts will reduce the workload for navigational officers when compared to using paper charts," Captain Gale says. "Voyage planning, route monitoring and positioning will be performed faster than with paper charts. Automatic updating of electronic charts will relieve officers of the tedious job of correcting charts and eliminate the risk of errors when making these changes" [5].

"Driven by the SOLAS-mandated transition to ECDIS, the adoption of digital navigation is the most visible sign of how navigational practices are being transformed. Additional tools are also playing an important role in supporting bridge teams and delivering accurate navigational information to the bridge [...] Vessels without an ENC service now represent a minority of the international fleet [...]The shipping world is embracing digital navigation with growing familiarity and confidence. For shipping companies that have installed ECDIS across their fleet and for mariners now serving on ECDIS-equipped ships, the new challenge is to develop and implement the practices and procedures that will able them to realize the full benefits for safe and efficient navigation," wrote Thomas Mellor at the UKHO [6].

"Technology is so advanced that ideas only previously dreamt up in fiction are now becoming a reality," Chief Officer Alexander McDonald says in the NI magazine *Seaways,* and that the dawn of ECDIS is upon us. Given that the primary function of ECDIS is to contribute to safe navigation, and that extremely sophisticated navigation technology exists, he has raised some very interesting questions: Why are we so reliant on ECDIS with so many flaws, weaknesses, and limitations? And more important, why do mariners tolerate them [7]?

"I have now sailed on paperless ships, each with a different ECDIS. That means that in the past years I have had to complete three different type-specific training courses. Why should we have to learn a totally new set of menus and buttons and terminology to perform day-to-day tasks? Why did the performance standards not limit the variation between different ECDIS systems?" Chief Officer McDonald asked, and continued: "These are complex computer programs and each one differs from the others. If equipment manufacturers will be upgrading their software in the near future and offering entirely new systems, navigators have to learn yet another type of operating program" [7].

The view of the chief officer has been discussed among navigators and he has found support and feedback. On the NI LinkedIn Forum, a specific discussion group was set up for identification of problems that could be faced regarding the IMO e-Navigation process. A UK-certified officer on watch (OOW), with experience on UK coastal and foreign-going ships commented that the systems are often unnecessarily complicated to operate as an end user:

"This is a major issue as many seafarers are not confident or competent in the use of computer systems. Lack of user confidence in either the users' own skill or the equipment is a serious problem as it may lead to the OOW doubting the information provided by the system. Lack of confidence or skill will undoubtedly lead to problems when planning a passage on the system and setting up the layers properly and effective monitoring" [8].

A master mariner said:

"In most discussions involving ECDIS, it usually gets a lot of criticism. Well, ECDIS has arrived. Period. We can sort out the wrinkles, but we cannot wish it away. If one gets overwhelmed by too many alarms and warnings of ECDIS, one should still be able to maintain one's situational awareness and navigate safely. In the old days of blind pilotage training, it was emphasized that while others were busy on radar and plots, the captain should keep his hand up and navigate. It is essential to be able to filter the flow of information and maneuver safely, rather than denigrate ECDIS that has substantially more advantages than disadvantages" [8].

But there are also other challenges on the bridge; the idea of a more or less standardized bridge layout being one thing and standardized equipment functions another.

"Personally I find it frustrating and annoying if when confronted by a new radar, say, I have to indulge in a kind of telepathic cooks tour of software to find what I need. Needless to say this also wastes time. Some are better than others and those with 'hard' button functions seem to be easier to use. More worrying is when the kit is in totally the wrong place. As a ship control centre the bridge has to satisfy many demands including navigation, communication, administration, cargo management, emergency response as well as generate its narrative in real time and proper task analyses is critical to generating the priorities needed for getting the man-machine interface right" [8].

These words are provided by a well-experienced MCA Master 200 Officer (certified for less than 200 tonnes) on the NI LinkedIn Forum. He continues: "When we talk about simplification and standardization, we also need to think about coherence and resilience. Future proofing is another matter entirely because civil sector technologies are racing ahead of the maritime sector's ability to respond. What makes me uncomfortable is when powerful lobbies impose prototype technologies before experience has ironed out the wrinkles," he commented [8].

In the Marsh Report of 2016, produced by Marsh's Global Marine Practice, it was reported that "over the past 12 months, ports around the world have heralded the arrival of ever-larger, record-breaking container ships, such as San Francisco welcoming the arrival of the CMA CMG Benjamin Franklin with 18.000 TEU (Twenty-foot Equivalent Unit) container capacity, and Gothenburg, Sweden, likewise doing the same with the MSC Maya with 19.225 TEU capacity [...] Commercial vessels of enormous proportions—in terms of length, width, and draught—are already or will soon be, plying waters around the world that have never before witnessed such vessels, or so many of them. From a commercial and economic perspective, this bodes well for maritime trade, as economies of scale should see transportation costs per unit of cargo reduce." In the same article, it is mentioned that in 2012, Andreas Tsavliris, then president of the International Salvage Union (ISU) said, "International business these days operates on the 'just-in-time' supply principle and container ships perform the function of floating warehouses." Hundreds of millions of dollars—even billions—worth of goods may be onboard.

The report continues, "Navigation routes, such as those leading to or from the Panama Canal, have been the same for many years, with commercial cargo vessels following tried- and tested pathways thorough the sea [...] However, accurate bathymetric data is inadequate or non-existent in large tracts of the world's oceans and seas, with large areas either un-surveyed or not re-checked since old lead-line soundings, measure in fathoms, were taken nearly a century ago [...] Many vessel operators have been viewing, with great interest, the increasingly viable Arctic routes between Asia and Europe as an alternative to the much longer (both in time and distance) routes via Singapore and the Suez Canal. However, the vessels that have, to date, successfully transited the Northern Sea Route (NSR) around Northern Russia have been relatively small in size. Marsh has already voiced concerns about the potential risks of larger vessels using this route with greater frequency [...giving] a loud warning about the possible consequences of deviating off planned routes [...In the worst case] a serious grounding or stranding, with all the perils of ship damage, crew endangerment, cargo loss and marine pollution [might be the result]" [9].

On the NI LinkedIn Forum, a captain with 37 years at sea and 15 years in command, commented on the article:

"I have transited the world normal routes in VLCCs/VLBCs (very large crude carrier/very large bulk carrier) with drafts of twenty meters plus. The sense of relief when clearing shallow water such as the English Channel or the Singapore Straits was tangible," he said.

A chief officer stated that he received voyage instructions to proceed to a port/terminal in the Philippines and that it was totally surprising that he received all largest-scale charts/ENCs but there was no information about the berth or the port. Even the Philippine local charts did not mention details of berth or port.

"It was clear after many messages flying here and there that the port and berth had been in use for the last, maybe, 8–9 years and still completely uncharted. When contacted pilots of that port they made the berth in approximately position by hand and clicked 'picture,' sending it to vessels concerned" [8].

The last financial crisis, fall in oil prices, and related activities ashore and offshore have over the last few years led to significantly reduced activities by many shipyards and equipment manufacturers. In order to further increase the competitiveness of this industry, initiatives on new technologies and processes have been introduced. However, voices have been raised, questioning whether existing regulations and performance standards limit further innovation, including reducing costs of systems and equipment without reducing the level of safety and security.

One of these initiatives, called Open Bridge and presented by a group of industrial stakeholders, was about multisupplier bridge systems. Creating new, innovative, and open-system architecture allows system suppliers and integrators to construct safer, more cost-effective, more user-friendly, and more flexible bridge systems together in a whole new way than what is possible today. Open and modular integration of bridge solutions might provide significantly more user-friendly and safe ships and ship bridges. Streamlining the integration process could reduce costs in equipment updates and navigational services in a ship's life cycle and be a completely new and powerful differentiation for suppliers, shipyards, and shipowners.

A key part here is that the bridge architecture can be realized cost-effectively based on today's legislation, but using a more futuristic interpretation of regulations. This requires close cooperation with stakeholders' responsibility for verifying that national regulations can be interpreted and enforced in such a way that the goal of effective comprehensive design is achieved while safety is maintained at the same time. The discussion has also focused on possible short-term and long-term changes in national and international regulations, which is necessary for further improvements of future integrated bridge solutions.

New integrated bridge systems from major suppliers have begun to be developed. Although they are more user-friendly and include a comprehensive design of bridge equipment, it is very challenging and often impossible to achieve this goal for bridge systems composed of equipment from different vendors, which is known as multisupplier bridge systems. User experience practices are defined by several parties independently. Due to a lack of coordinated development of multisupplier bridge solutions, three main issues occur:

1. The work of bridge installation is fragmented and not user-friendly.
2. The costs of developing user interfaces are very high with little reuse of solutions across vendors.
3. The threshold of integrating new and innovative software on the bridge is very high. In addition, these types of bridge systems are costly and labor-intensive to install and test.

The bridge architecture facilitates open integration by creating common specifications of user-centered design principles and will integrate this with an accompanying and enabling IT infrastructure. One part is related to the user and safe operation, while the other part is related to IT solutions and technical standards.

Total integration on board doesn't necessarily provide a better solution. A case in point is the cruise liner the MS *Royal Majesty,* which was fitted with a fully integrated bridge system but was unable to let the system know that the GPS antenna was broken, yet the GPS was still sending a position that was invalid. The resulting accident was due to a mixture of equipment problems and poor crew awareness. More about this accident is described in Chapter 5.

The examples above show that navigators face many challenges on board the bridge when navigating safely and efficiently. Lack of built-in functionalities to ensure that navigational equipment is displaying the correct information without manually cross checking with other equipment, lack of situational awareness, lack of coordinated display of information, and differences in the displays and controls provide a poor working environment. Furthermore, navigational information received via communication equipment is currently printed and has to be read, analyzed, and transferred where necessary rather than being displayed directly on the navigational systems, in addition to lack of navigation displays in standardized manner. These issues are all of central importance.

2.2 Ship Reporting and Shore-Based Services

Captain Christian E. Rørbeck from Denmark, in command at the time of the largest container ships ever built, was invited in 2013 to provide a presentation at the IMO's Facilitation Committee (FAL Committee) in London about the administrative burdens faced by ships on a daily basis around the world.

According to *IMO News* [10], his specific examples of administrative burdens for a container ship in service were an eye-opener for IMO delegates. The focus of the presentation was the IMO FAL forms, which are standardized documents that public authorities can demand of a ship, and the countless local variants of these in connection with port calls in 17 randomly selected ports around the world. According to the captain, the problem is not the port and prearrival documents themselves; the problem lies in the amount of paperwork and the fact that each port has its own requirements regarding paperwork to be submitted both prior and on arrival. In his own estimate, he spends 80% of his work time on paperwork.

Many ports have requirements for prearrival information. The time limit for sending this information is not the same globally, which means that some ports want the information 24 hours in advance, others 48 hours in advance, and still others 72 hours in advance. It was noted by Captain Rørbeck that some port authorities do not report back whether they have received the submitted information. In addition, the shipping companies' agents handle the reporting in some ports, whereas the master must communicate directly with the port authorities in others. This goes for both port and prearrival documents.

Upon arrival, each port also has its own requirements regarding the paperwork that must be submitted. Some of the information may already have been submitted electronically, but on port arrival the same documents must be printed, signed, and stamped.

Several European countries have joined the Schengen agreement. There are normally no border controls when transiting from one country to another within EU's Schengen area. However, every time a ship calls on a port, all the paperwork has to be prepared and submitted. Captain Rørbeck showed that, for a ship traveling in the Schengen area, a total of 80 documents must be prepared and 42 documents must be sent in advance [10].

However, in the EU there is a drive to ensure that ship reporting is done only once and that the shore authority receiving the information distributes the relevant data to the appropriate bodies, which are called Maritime Single Window (MSW), by using the SafeSeaNet (SSN) system.

Directive 2010/65/EU of the EU is about reporting formalities for ships arriving in and/or departing from ports of EU member states and how the information procedures can be simplified and harmonized, including how the information could be gathered more effectively. EU countries should accept electronic reports via a single window as soon as possible. The Directive states, "The single window will be the place where all information is reported once and made available to various competent authorities and the EU countries. EU countries must ensure that information received in accordance with reporting formalities is made available in their national SafeSeaNet systems and make available parts of such information to other EU countries via the SafeSeaNet system. EU countries shall accept [IMO] FAL forms [...] for the fulfillment of reporting formalities" [11].

An example here is the protection of the marine environment that can be enhanced by reducing discharges into the sea of ship-generated waste and the remnants of cargo material.

Directive 2000/59/EC of the European Parliament and of the Council of November 27, 2000 on port reception facilities for ship-generated waste and cargo residues, now linked to Directive 2010/65/EU, enhances the availability and use of port reception facilities for ship-generated waste and cargo residues. It also sets out a regime of enforcement, including a system for inspections and for the exchange of information. The directive covers ships, regardless of their flag, which arrive at port of a EU country.

Captains of ships bound for an EU port "are required to notify certain information, in particular:

- The date and the last port in which ship-generated waste was delivered;
- The types and amounts of waste and residues to be delivered and/or remaining on board and the percentage of maximum storage capacity" [12].

Captain Jerzy Najmola from Poland has been following with great attention any initiative aimed at standardization and simplification of administrative procedures in the maritime sector and therefore he appreciated the MSW initiatives. As an active captain in international shipping, he has analyzed such initiatives from a master's perspective, considering that the ship is only one part of the data exchange chain between ship and competent authorities. However, his earlier experiences in this matter told him to be extremely careful and far from optimistic. In a personal interview, he stated, "Many of the earlier attempts to introduce uniform standards resulted in a significant and senseless increase of administrative burdens for ship's Masters."

As an example, according to the captain, he could talk about 50 different forms in Europe, showing the distribution and diversity of forms on waste reporting. He has created a drawing, shown in Figure 2.1, based on his own experience on the situation in 2015 just before the planned deadline of June 2015 for the project. Although the deadline was not met by some EU member states he would like to underline that he looks at the MSW project in Europe with great hope but also with fear.

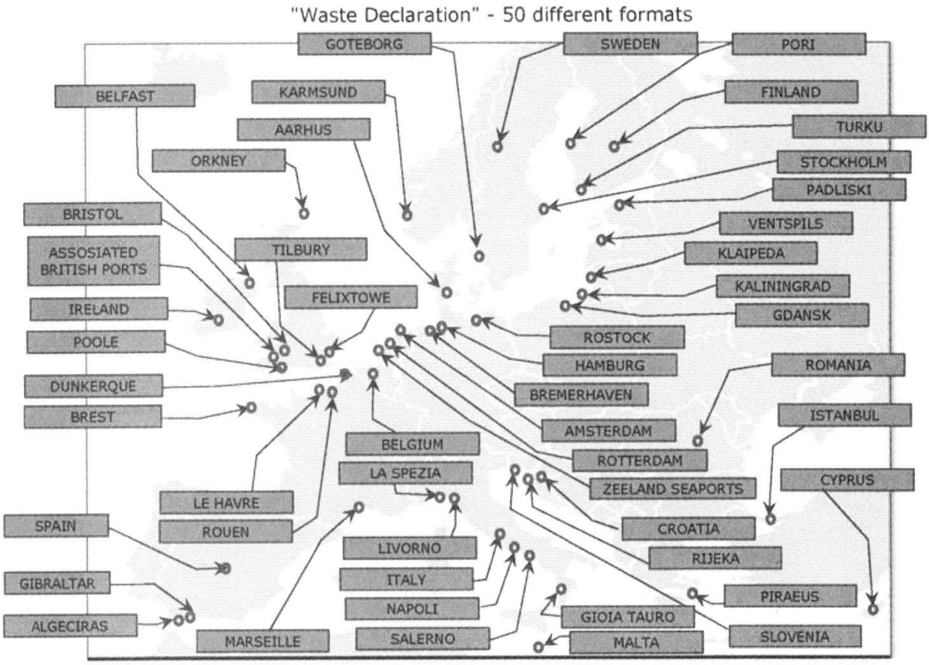

Figure 2.1 Captain Jerzy Najmola's own experience of the number of different waste reporting forms in Europe in 2015, which total 50.

"I am perfectly aware that the Captain will have to fill in and submit an electronic information form to the relevant authorities, but this will only be an attempt to gather all data required in all EU States because he feels they are all different. This is an extremely ambitious project. That is exactly why it should be designed in an absolutely perfect way, leaving no doubts even for inexperienced users, regardless of the time frame constraints. It is ambitious and time-consuming project as well and therefore there should not be any doubt or ambiguity from the beginning."

He feels that each and every master expects that the modern system of data exchange will be easy to use, efficient, and user-friendly. He has sailed worldwide to all continents and has gathered data on the different forms used for reporting. Of the 86 ports called at by his ship, the data shown in Table 2.1 was obtained according to his own experience.

Table 2.1
Captain Jerzy Najmola's Data on Different Forms Worldwide Used for Ship Reporting in 2015

Continent	Number of Ports Analyzed	Total Number of Documents Analyzed	Number of FAL Forms Used	Result in % of FAL Forms Used
Europe	26	409	121	30%
Asia	29	526	74	14%
North America	11	156	15	10%
Sout America	15	354	47	13%
Africa	5	98	19	19%
Total	86	1,543	276	18%

This table shows that only 18% of the forms he had to fill out were standardized FAL forms, ship reporting forms defined by the IMO.

At an international e-Navigation conference in Singapore in 2014, Director-General of the Norwegian Coastal Administration, Kirsti L. Slotsvik, presented Project SSN Norway, which was about building a complete MSW solution as part of e-Navigation implementation. The concept was developed in accordance with e-Navigation solution 2.

Instead of providing a ship's information to many different agencies in a port, such as customs, immigration police, and others, the information is entered only once on the ship and it can be cleared by one electronic submission that is automatically routed to the relevant authorities. In order to establish a complete MSW solution (see Figure 2.2), the following information from a ship would be included [13]:

- Level 1: 7 IMO FAL forms
- Level 2: European legislation (directives)
- Level 3: National legislation and requirements

SafeSeaNet Norway facilitates a complete MSW solution to ships arriving at Norwegian ports and is cost effective, reducing huge administrative burdens onboard and ashore. The next step of

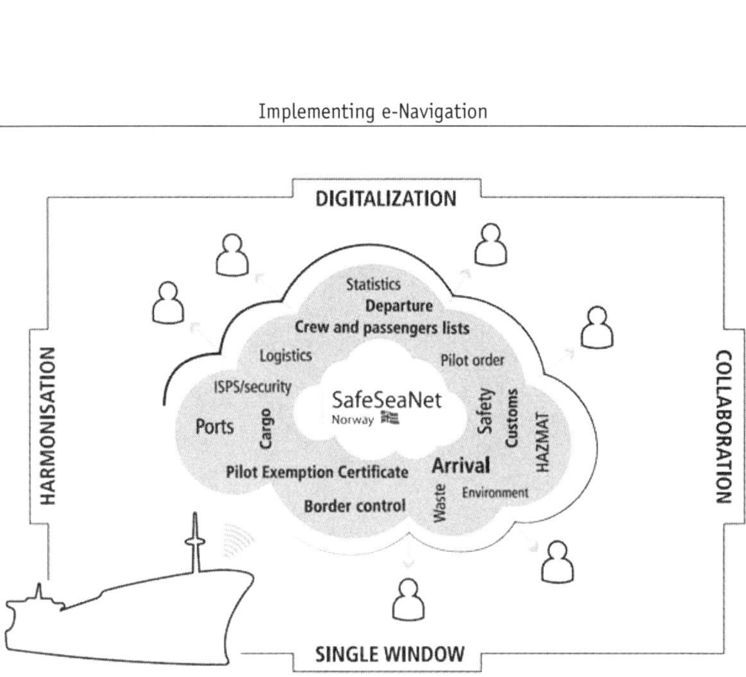

Figure 2.2 SafeSeaNet Norway provides harmonized digital information. (Illustration: Norwegian Coastal Administration/Jarle Hauge.)

the project is to establish a system for automated or semiautomated reporting. You will find more about this in Chapter 4.

Furthermore, mandatory requirements for the electronic exchange of information on cargo, crew, and passengers were adopted in 2016 as part of a revised and modernized annex to the IMO FAL Convention, which aims to harmonize procedures for a ship's arrival, stay, and departure from a port. The new standard relating to obligation of public authorities to establish systems for the electronic exchange of information within a period of three years after the adoption of the amendments, is among important changes in the revised annex. A new recommended practice also encourages the use of the single window concept, to enable all the information required by public authorities in connection with the arrival, stay, and departure of ships, persons, and cargo, to be submitted via a single portal without duplication.

In light of the ongoing work by the international community to mitigate climate change, in 2016 the IMO's Marine Environment Protection Committee (MEPC) approved mandatory requirements for ships to record and report their fuel consumption. Under the system, ships of 5,000 gross tonnage and above will be required to collect consumption data for each type of fuel they use, as well as other additional specified data. The aggregated data will be reported to the flag state after the end of each calendar year and the flag state, having determined that the data has been reported in accordance with the requirements, will issue a Statement of Compliance to the ship [14].

It is also interesting to note that the IMO has defined tasks and timelines to reduce greenhouse gas (GHG) emissions from ships.

2.3 Communications and Interoperability

The U.S. Department of Homeland Security reports, "Although telecommunications technology is improving quickly, people at sea do not have access to the same telecommunications infrastructure people ashore have. Like people ashore, mariners need to access international shore telephone and data public switched networks. Additionally, they need to access many maritime specific communications listed below. Mariners need to be able to

- Communicate with other ships of any size or nationality;
- Receive and send urgent maritime safety information (MSI);
- Send or receive distress alerts in an emergency to or from rescue coordination centers ashore and nearby ships anywhere in the world" [15].

In 1988, the Global Maritime Distress and Safety System (GMDSS) was established to increase safety and make it easier to rescue ships in distress. Chapter IV of SOLAS was introduced to define required GMDSS functionalities and supporting communications equipment, and is applicable to all passenger vessels and cargo vessels above 300 gross tonnage that are engaged in international voyages.

Most navigational information received via communications equipment is currently printed and has to be read, analyzed, and transferred where necessary rather than being displayed on navigational systems. Broadcast and reception of MSI by means of direct printing is an important part of the GMDSS. On most ships, MSI information received via communication equipment such as

NAVTEX, which is an international automated medium frequency direct-printing service for delivery of navigational and meteorological warnings and forecasts as well as urgent maritime safety information to ships, and INMARSAT-C, a two-way packet data service operated by the telecommunications company Inmarsat, which operates between mobile earth stations (MESs) and land earth stations (LES), are either displayed on separate screens or printed on a scroll of paper [16].

Communication used as a navigational aid and for mandatory reporting to maritime authorities includes two-way voice communication by radio or satellite, AIS, LRIT, VTS service and port arrival notifications (FAL requirements), and different ship reporting schemes for port states (e.g., related to dangerous goods and marine pollution). As with GMDSS safety communications, these applications are mandated by regulations and do not require broadband data connection.

Looking at state-of-the-art communication, DNV GL, an international accredited registrar and classification society, said that maritime authorities do not drive the developments of maritime broadband communications. This lack of drive and technology push is a barrier, and probably a key reason for the lack of data services in the VHF, MF, and HF bands. The bandwidth available in these bands is already severely limited, so any new data service will remain narrowband compared with the alternatives such as satellite and terrestrial mobile. Other barriers include service coverage, which is limited to a distance from the shore base stations such as VHF and MF, and the service, which is limited for long-range connections like HF [16].

A chief officer, providing an example on the NI LinkedIn forum on improper use of VHF Channel-16 said, "We all have heard long conversations going on between two absent minded people on VHF Channel-16, where someone said stop which resulted in abuse which makes a mockery of a protected radio channel. Sometimes the height of nuisance is when a certified watchkeeper puts a song on VHF Channel-16."

"These actions are worldwide known to almost all mariners and they themselves know it's illegal and unprofessional. Now when so many technological developments are being conducted why isn't this aspect raised in IMO and other authorities to make strong legislations and equip port VTS/VTIS (vessel traffic service/vessel traffic information services) which can detect such nuisance by tracking continuous signal transmission tracking," he proposed.

In this respect, a GMDSS expert seriously emphasized that VHF is one of the major means for communication at sea. VHF Channel-16 is designed especially for distress, urgency, and safety communications. For routine communication, it is allowed to use VHF Channel-16 as an initial calling channel, and after an initial communication has been established, the call must be switched to one working channel. Improper use of VHF Channel-16 could be dangerous for safety of navigation. Poor ship management, lack of means for monitoring VHF equipment, and poor competency of seafarers might be considered as the main contributory causes leading to improper occupation of marine VHF Channel-16 [8].

Technology development on legacy radio systems is generally limited. However, VDES is a promising initiative to provide a digital data service based on the existing terrestrial VHF radio infrastructure. IALA has proposed VDES with the aim of offloading the VHF Data Link (VDL) of AIS and creating a new channel for data exchange to support e-Navigation applications. This defines more spectrum-efficient modern techniques and higher data rates, which are 32 times the speed and eight times the efficiency of existing AIS channels.

DNV GL said in the report *DNV GL Strategic Research & Innovation Position Paper 04-2015 Ship Connectivity* that a different type of communication driver relates to the main operational purpose of the vessel. This type of communications application is not driven by regulations, but rather deployed voluntarily in order to optimize operations, thereby saving costs to the shipowner or increasing service quality for the charterer. Such operational applications could be

- Cargo logistics and monitoring applications;
- Route planning and energy efficiency applications;
- Administrative communication between the vessel and the shipowner's HQ;
- Upload of gathered data such as seismic or environmental data.

Operational communications are a relatively modern driver that has emerged over the last decade through the age of digitization and satellite communications. Operational applications have been a key driver for shipowners to invest in new broadband communication capabilities on board their ships.

It is also interesting to note that welfare and entertainment for crew and passengers has been one of the strongest drivers for investment in advanced communications equipment in recent years. In general, the maritime sector is in the midst of a connectivity revolution, in which more and more vessels will become broadband capable.

Terrestrial-based mobile communications have, however, already enjoyed enormous growth for land users over the last two decades, and base stations along the coast handle maritime traffic. But terrestrial systems have some natural coverage limitations, which means that they are unable to support vessels during the deep-sea part of their voyages. Another barrier is traffic costs [17].

The rapid development of digital technology involved with equipment and systems should, as far as practical, be forward- and backward-compatible and support integration wherever possible. Any communications systems used must be able to deliver appropriate digital information. The highest level of interoperability should be sought where practicable. Any impacts affecting existing systems must be based on technical standards and protocols for data structure, technology, and bandwidth and frequency allocations.

So, what is the current overarching situation and ongoing work regarding further development on maritime communications systems?

The two United Nations-chartered organizations, the IMO and the International Telecommunication Union (ITU), are responsible for defining and regulating maritime telecommunications systems.

The ITU is the United Nations' specialized agency for ICTs. The 2016 edition of the ITU's annual global ICT regulatory report, *Trends in Telecommunication Reform 2016,* charted and analyzed the challenges and opportunities facing today's ICT regulators as services proliferate, platforms converge, and network operators ready their infrastructure for the next round of data-intensive technologies, from 5G mobile to IoT, according to the report [18].

New ICT goods and services are bringing with them enormous social and economic disruption. But while there are many benefits to be reaped, the report highlighted regulatory challenges that must be addressed to preserve the level playing field considered essential to competition and innovation. With the theme of exploring regulatory incentives to achieve digital opportunities, the report emphasized the growing importance of flexible, light-touch, technology-neutral regulatory principles that stimulate market growth while protecting the rights of consumers and encouraging new entrants.

"Information and communication technologies are now all-pervasive and will be central to helping the world meet all 17 Sustainable Development Goals. The role of ICT regulators in creating an enabling environment for ICT growth and development has never been more critical," said Houlin Zhao, ITU Secretary-General in an ITU press release. "ITU's annual *Trends in Telecommunication Reform* report supports regulators around the world to put the right policies in place for their national markets" [18].

The press release also referred to Director Brahima Sanou of the ITU's Telecommunication Development Bureau. He said, "This 16th annual edition of *Trends in Telecommunication Reform* focuses on regulatory incentives to achieve digital opportunities which I believe upholds the ultimate goal of regulators: serving consumers" [18].

e-Navigation will need interoperability, and the report stated that the concept of interoperability is much broader than just technical compatibility, with implications across all four key levels: technological, data, human, and institutional. According to the IMO, "The highest level of interoperability between e-Navigation and external systems should be sought where practicable" [19].

"Systems can increase interoperability by providing greater opportunities for technical interconnection; being more open about the types of systems and services that can interconnect; supporting a greater variety of data; and by making it easier for humans to leverage the interconnections a system that has more points of access allows more types of systems to connect, processes data with fewer limitations, increases potential attack vectors and creates more opportunities for nefarious actors to exploit data or to inject bad code" [18].

The World Radiocommunication Conference in Geneva in 2012 (WRC-12) provided a Resolution (Resolution 359 (WRC-12)), "Consideration of regulatory provisions for modernization of the Global Maritime Distress and Safety System (GMDSS) and studies related to e-Navigation to improve efficiency and introduce bands for new digital technology, including reviewing the regulatory provisions and spectrum allocations for use by maritime safety systems for ships and ports." It was recognized that advanced maritime communication systems may support the implementation of GMDSS modernization and e-Navigation, and that IMO efforts to implement GMDSS modernization and e-Navigation may require a review of the radio regulations to accommodate advanced maritime communication systems. According to ITU Resolution 810 (WRC-15), which is about the preliminary agenda for the 2023 World Radiocommunication Conference, consideration of "possible spectrum needs and regulatory actions to support the GMDSS modernization and the implementation of e-Navigation" are on the agenda [20].

It is obvious that maritime telecommunications system must be internationally interoperable. Bringing new telecommunications technology to mariners might be challenging, since to be interoperable, the technology must be affordable, acceptable, and available to most ships and maritime countries.

2.4 Challenges in VTS and Port Areas

VTS is a service implemented by a competent authority that is designed to improve the safety and efficiency of vessel traffic and to protect the environment. The service should have the capability to interact with traffic and to respond to traffic situations developing in the VTS area. SOLAS Chapter V, Regulation 12 states that the use of VTS may only be made mandatory in sea areas within the territorial seas of a coastal state [21].

IALAs *VTS Manual 2012* [22] underlined that traffic density and structure, navigation hazards, local climate, topography, environmental requirements, commercial aspects, and the extent of a VTS area sets the requirements for a VTS. These factors will have substantial impact on life cycle costs of a VTS and the acquisition of VTS equipment. Equipment may include

- Communications;
- VTS radar system;
- AIS;
- Closed-circuit TV (CCTV) camera;
- RDF;
- Hydrometeo equipment;
- VTS data system.

The required features, and in particular, the need for coverage by sensors such as radar, should be determined by an assessment of a service to be provided, the safety level to be achieved, and the user requirements of the VTS system.

A VTS can monitor traffic and provide information services and navigation assistance to vessels where necessary and if desired. Many areas are under national or local restrictions and the VTS represents an authority, approving clearance for movements in the area, which might be a coastal or port area (see Figure 2.3).

If a vessel on arrival in a VTS area needs to seek permission prior to entering, the captain must submit an application for clearance orally over a VTS VHF channel. Upon the submission of such an application, there will often be challenges of a different nature. VTS operators have reported that language or communication systems challenges with noise and VHF cover area are often the reason that officers have trouble explaining their planned itinerary or other relevant information.

Misunderstandings can easily arise especially when the communicating parties don't speak the same language and are both speaking a third language, such as English. These include ship-ship or ship-VTS communication.

VTS operators have experienced that some navigators do not understand what is being communicated because of language difficulties. This still applies when a standardized language is used, such as the IMO Standard Marine Communication Phrases (SMCP).

Navigational and safety communications from ship to shore and vice versa, ship to ship, and on board ships must be precise, simple, and unambiguous, in order to avoid confusion and error. As such, there is a need to standardize the language used. This is

Figure 2.3 Kvitsoy VTS on the western part of Norway. (Photos: Norwegian Coastal Administration/Haakon Nordvik.)

of particular importance in light of the increasing number of internationally trading vessels with crews speaking many different languages, since problems of communication may cause misunderstandings leading to danger to the vessel, the people on board, and the environment. The IMO SMCP is designed to overcome this problem but in reality sometimes has difficulties [23].

When sending the same message to multiple vessels, a VTS center needs to call each vessel in turn in order to convey the same or nearly the same message every time. Examples of this are areas of flotsam, which can be floating wreckage of a ship or its cargo or search and rescue (SAR) operations.

The establishment of a more efficient SAR function in the future would largely depend on an automated data network connecting all stakeholders in SAR intervention, include improved communication between maritime rescue coordination centers (MRCCs) and shore-based, sea-based, and air-based entities, and the fact that modern technology should be capable of providing all information relevant for SAR purposes both to and from SAR services.

Sufficient landside capacity to keep cargo moving is essential for liner vessels to maintain their schedules. Supply chains served by liner vessels in the twenty-first century are no longer port-to-port. All cargo carried on liner ships must be able to discharge the vessel in a timely manner at port facilities around the world.

This cargo will also have to move via truck or rail on its journey from origin to destination. An intermodal network is comprised of ships, trains, planes, and trucks, including the surface over which they move and the connections or transfer points between the modes, often referred to as intermodal connectors. Service disruption or insufficient capacity anywhere in the network could result in shipment delays and increased cost.

The efficiency of international cargo transportation systems is an important economic issue; however, creating safer waterways by reducing ship accidents and environmental incidents is of vital importance. These issues are paid particular attention by mariners.

A safe and effective logistics system in port areas often requires detailed preparations. An example is the MSC *Splendida,* which docked with more than 4,600 passengers and crew on board in Dublin Port. *The Irish Mirror* informed its readers about how the world's 11th longest cruise ship, the MSC *Splendida,* reversed its way up the mouth of the Liffey and into Dublin Port. The arrival of the biggest ship to ever dock in the port required a good deal of planning. "The preparation would have started a number of weeks ago when we were approached by MSC (MSC Cruises) and asked would we be able to bring such a big ship in," said Dublin Port chief executive Eamonn O'Reilly.

"The captain, Gieuseppe Maresca, would have done simulations at the National Maritime College of Ireland with our harbour master David Dignam and with our pilots to see how you maneuver such a large ship into a relatively confined port like Dublin."

Being too big to turn within those confines, the 140,000-ton liner had to enter Alexandra Basin stern first. "Everything went very smoothly," said Mr. O'Reilly [24].

An operation like this requires close cooperation between the captain and crew and pilots and the local port authority, recognizing the role of the practice of good seamanship, the provision of suitable training, and the use of procedures and modern technology.

Furthermore, in light of advances in technology over the last decades, which have been substantial and rapid, many have raised the question how ships' navigators and maritime shore-based stakeholders who are a part of a national or international transportation systems could contribute to even more safer, more secure, more reliable, more efficient, and more integrated transport.

In this respect, work at the IMO and other international organizations has for a long time been in progress regarding navigational equipment, systems, displays, integrated bridge systems, ship reporting, and shore-based services, in addition to preparing for advanced maritime communication systems that can support the implementation of e-Navigation. This particular voyage started with identification of user needs and is moving toward the final destination—the practical use of e-Navigation. You will find more about this process in Chapter 3.

References

[1] MONA LISA 2.0: Securing the Chain by Intelligence at Sea (2013), available at http://www.sjofartsverket.se/pages/36039/MONALISA%202_0%20web.pdf, accessed May 10, 2017.

[2] Regulation 19, "Chapter V, Safety of Navigation," in *Annex to the International Convention for the Safety of Life at Sea (SOLAS V)*, available at http://solasv.mcga.gov.uk/regulations/regulation19.htm, or from the IMO, accessed May 10, 2017.

[3] American Pilots' Association, *Portable Pilots' Units: A Best Practices Summary*, October, 27, 2016 [online], available at http://www.americanpilots.org/PPUBestPracticesAPA_%2027Oct2016.pdf, accessed May 10, 2017.

[4] United Kingdom Hydrographics Office, "Over Half of Internationally Trading Fleet Now ECDIS Ready," December 7, 2015 [online], available at https://www.admiralty.co.uk/AdmiraltyPressReleasesLibrary/071215%20-%20UKHO%20Press%20Release%20-%20Over%20half%20of%20internationally%20trading%20fleet%20now%20ECDIS%20ready.pdf, accessed May 10, 2017.

[5] Gale, H., ECDIS Transition Document 09/02/2009 [online], available at https://www.hydro-international.com/content/news/ecdis-transition-document, accessed May 10, 2017.

[6] Mellor, T., "ECDIS: The Quiet Revolution in Maritime Navigation," May 19, 2016 [online], available at https://www.marinelink.com/news/revolution-navigation409963, accessed May 10, 2017,

[7] McDonald, A., "Designing an ECDIS That's Fit for Purpose," *Seaways*, February 2016.

[8] The Nautical Institute, NI LinkedIn forum, which is an open forum and does not necessarily constitute the views of the Institute. This is a private group.

[9] Marsh Report, March 2016, republished as "Hydrographic Concerns for Modern-Day Large Vessels," October 14, 2016 [online], available at http://www.ee.co.za/article/hydrographic-concerns-modern-day-large-vessels.html.

[10] Rørbeck, C., *IMO News*, Issue 3, 2013.

[11] Directive 2010/65/EU [online], available at http://eur-lex.europa.eu/legal-content/LV/TXT/?uri=URISERV:tr0047, accessed May 10, 2017.

[12] Directive 2000/59/EC [online], available at http://eur-lex.europa.eu/legal-content/EN/TXT/?uri=LEGISSUM:l24199, accessed May 10, 2017.

[13] Slotsvik, K., The Norwegian Single Window Solution [online], available at http://www.mpa.gov.sg/web/wcm/connect/www/72bf3bd7-f1a8-4ab0-8c5c-fdf25e8c09d1/slotsvik_norwegian_single_window.pdf?MOD=AJPERES, accessed May 10, 2017.

[14] *IMO News*, Issue 2, 2016.

[15] U.S. Department of Homeland Security, United States Coast Guard, The Navigation Center, Maritime Communications, [online], available at https://www.navcen.uscg.gov/?pageName=maritimeTelecomms, accessed May 10, 2017.

[16] IMO, MSC/95/19/xx, March 3, 2015 [online], available at http://www.ics-shipping.org/docs/default-source/Submissions/IMO/work-programme-implementing-e-Navigation-to-enhance-the-safety-of-navigation-and-protection-of-the-marine-environment.pdf, accessed May 20, 2017.

[17] DNV GL, *Strategic Research & Innovation Position Paper, Ship Connectivity 04-2015.*

[18] ITU, *Trends in Telecommunication Reform 2016* [online], available at www. http://www.itu.int/net/pressoffice/press_releases/2016/12.aspx#.WRMJ6U2wd1s, accessed May 10, 2017.

[19] IMO document, NAV54/25 Annex 12.

[20] ITU, Resolution 810, WRC-15.

[21] Regulation 12, Chapter V, Vessel Traffic Services, *Annex to the International Convention for the Safety of Life at Sea (SOLAS V)*, accessible at https://mcanet.mcga.gov.uk/public/c4/solas/solas_v/Regulations/regulation12.htm, accessed May 10, 2017.

[22] IALA, *VTS Manual*, 2012 Edition 5.

[23] IMO document, NAV 46/INF.4.

[24] O'Reilly, E., in "World's 11th Longest Cruise Ship Arrives in Dublin Port," *The Irish Times*, May 11, 2015, by D. Griffin and É Ó Caollaí, [online], available at http://www.irishtimes.com/news/ireland/irish-news/world-s-11th-longest-cruise-ship-arrives-in-dublin-port-1.2207709, accessed May 10, 2017.

3

Performance Gaps

3.1 Identifying User Needs

It has been a decade since the IMO took the first steps to "...develop a strategic vision for the utilization of existing and new navigational tools, in particular electronic tools, in a holistic and systematic manner." At that time, in 2006, the IMO gave the nod to a proposal from seven Member States, who requested work commence to develop an e-Navigation strategy [1].

A year earlier, in 2005, in a move to solicit wider support for an e-Navigation strategy, Brian Wadsworth, then director at the UK Department for Transport, developed an orientation paper mentioned in Chapter 1 that argued, "We have an important window of opportunity to make marine navigation easier and to reduce navigational errors, with their attendant toll of accidents, loss of life, injury and environmental damage" [2]. The rest is history. The IMO and other international organizations embarked on this work and developed an e-Navigation strategy in 2008, and then a plan to implement the strategy in 2014.

The IMO agreed that the implementation of e-Navigation should be a phased iterative process of continuous development.

The first step was identification of users and their requirements or needs. The next step was identification of groups or functions or services, or a combination of these two categories, needed to meet these navigational needs (see Figure 3.1).

A definition of the integrated e-Navigation system architecture and concept of operations should be based on consolidation of the identified user needs across the entire range of users. The functional and technical architecture should include hardware, data, information, communications, and software.

There was a common understanding from the outset of e-Navigation development that the identification of performance gaps should relate to relevant user needs. A number of papers on user needs were presented to the IMO and became the basis for the description of the preliminary shipboard user needs. Potential needs of ship- and shore-based users of e-Navigation were also presented [3].

In total, 24 user needs were identified and approved [3]. Fourteen of the user needs were identified onboard in the following categories:

Human-machine interface issues:

- Improved ergonomics;

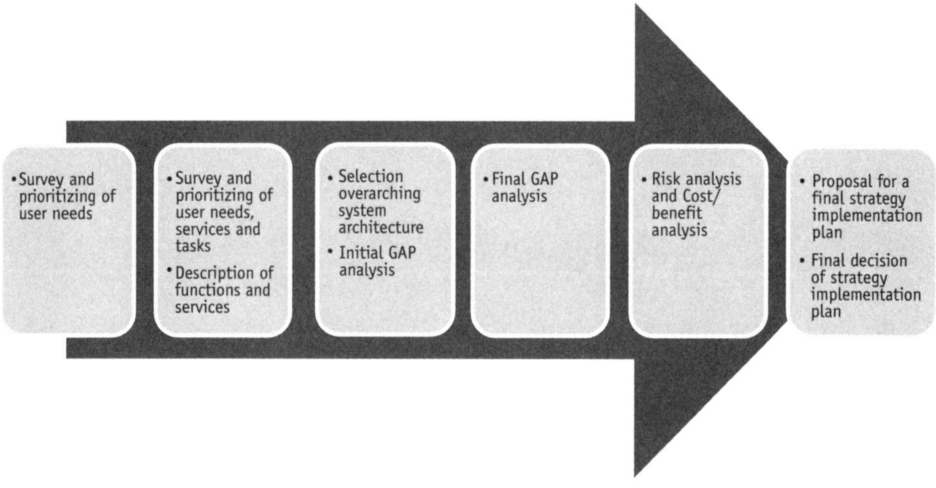

Figure 3.1 Step-by-step roadmap of the IMOs e-Navigation development. (Source: IMO document MSC 86/23/4 .)

- Standard interface;
- Familiarization requirements;
- User-selectable presentation of information received via communication equipment;
- MSI;
- Alert management;
- Indication of reliability;
- Operational issues;
- Improved reliability;
- Standardized and automated reporting;
- Improved target detection;
- Guard zones;
- Reduction of administrative burden and increase use of electronic documentation;
- Automated updating of baseline data and documents;
- Effective and robust communications.

Six shore-based user needs were identified

- Collection of information;
- Management of information;
- Provision of information to vessels;
- Quality assurance;
- Shore-to-shore information exchange;
- Effective and robust communications.

Four SAR authority user needs were identified

- Access to relevant information contained within e-Navigation domain;
- Effective communication and information sharing;
- Priority for distress communications;
- Access to the details of all relevant onboard communication equipment and capabilities.

The main purpose of the needs analysis is user satisfaction, which aimed to understand the user's needs. As it focuses on the needs of the human, the needs analysis is not limited to addressing the requirements of software, but can be applied to any domain, such as products or services. Although it is not a business development tool, it can be used to help in the development of a business case. Information gathered when establishing context of use can help scenarios that provide realistic examples of future use.

The next step, the identification of the groups of functions or services needed to meet the primary navigational needs, was based on a structured, systematic, and traceable methodology relating the functions to tangible operational benefits. It was repeatedly emphasized that the development of e-Navigation should be user-driven and not technology-driven. However, perceived user needs today can be hugely different from those of a mariner several years ago, not least because of the advances of technology—which changes the perception of detailed user needs.

It is natural that the priorities may be different for IMO member states, shore-side organizations, shipowners, and mariner organizations. For a successful result, it was necessary that a close and confidant collaboration was further developed between the various stakeholders involved.

It was also agreed that there should be harmonization between the shipboard and shore-based systems and procedures, and coordination of inputs into e-Navigation development from shipboard and shore-based users and other relevant bodies. In particular, the agreed-on shore-based identified user needs required further development, and there also was a need to address effective ship-shore interoperability.

It was recommended that international organizations such as IHO, IALA, and others provide input to the IMO development of e-Navigation, and that IALA committees should be invited to contribute.

At the same time, a list of preliminary shipboard user needs was agreed on [4]. Although not an exhaustive list, it was considered to be comprised of realistic operational goals, which in turn would form the basis of an initial gap analysis.

The agreed-on shipboard user needs fell into three main categories: operational improvements, human-machine interface (HMI), and reliability:

Operational improvements included the need for user-friendly presentation of information, better alarm management, standardized and automated reporting, improved target detection, and better use of guard zones. The emphasis for these needs was to make better use of technology to reduce distraction and offer better decision-making support to the navigator. These issues were considered particularly important to fulfill the IMO e-Navigation strategy for reducing single-person errors.

HMI issues included the need for better ergonomics in general, but specifically the use of more standard interfaces for bridge equipment to improve operational effectiveness and to reduce onboard familiarization challenges. Mariners also requested that there should be better graphical representation of MSI on their primary navigation displays and that known inaccuracies, for example from electronic positioning systems or hydrographic surveys, be effectively represented.

Reliability was a major issue for mariners, as they had seen an increased use of electronic systems onboard with varying degrees of reliability, due to software, hardware, data quality, power supply, or even installation or repair standards. Electronic equipment is vastly more reliable than it was in the past, as there is much more of it now and it performs far more complicated tasks. However, as discussed in Chapter 1, digitalization and the use of cyberspace increases the number of potential threats, but also provides the potential for greater automatic vigilance on existing threats.

Another challenge was that e-Navigation HMIs should be consistent in their layout and operation, so that users can easily move from one work station to another and between ship systems. This must apply to both shore users and ship users.

A review of the preliminary detailed shipboard user needs was carried out next. The inputs, especially those related to the interaction between ship and shore, clearly identified the need for a greater efficiency in both data and voice communication.

There had also been a process to develop detailed shore-based user needs, taking into account inputs provided by IALA and IHO.

User needs concerning SAR at sea were also included. Some IMO member states conducted in-depth studies concerning a significant number of users on shore and on board. Language skills, reporting, information to be displayed, use of AIS, S-Mode, terrestrial backup, and communication were documented as central issues of e-Navigation user needs.

e-Navigation systems should be designed specifically to reduce single-person errors, particularly on board. This also generated discussions on topics such as, Who is going to be on the ship in the future, what do we want those people to do, and what should their qualifications be to satisfy the requirements of e-Navigation.

The discussion on the issue of the future role of the navigator resulted in two scenarios [5].

The first scenario was of the navigating navigator. "This is a scenario where the monitoring equipment is kept relatively traditional on board and ashore. The navigators' own skills will still be essential to the safe navigation of the ship, and the bridge team will be the main backup to the safe functioning of the ship. This will have to be reflected in the principles of the training and certificates required. There is, however, also a question of whether one should emphasize assessment of the de facto skills and competencies, or alternatively if assessment should include a more formal documentation of having fulfilled training programs" [6].

The second scenario was of the monitoring navigator. "In this scenario the data solutions and monitoring equipment are much more sophisticated. The navigator will have to rely more heavily on automated processes, standardized and harmonized procedures and equipment. Data structure, displays and services will have to be interoperable. A main task will be to monitor the system displays and the indicators of the system's health and resilience. This scenario will include an even closer cooperation with organizations ashore to assist a safe voyage from berth to berth. A consequence of this scenario is that the competence of the seafaring professional could be affected, and there would be implications for the development of the training, education and required competencies for seafarer certificates" [6].

It was underscored that the navigator's own skills would remain essential for the safe navigation of the ship, and the bridge team would be the main backup for the safe functioning of the ship. It would not be advisable to be totally reliant on systems where the navigator only monitors the system displays and the indicators of the system's normal functionality or resilience. It was also expressed that this should not be an either/or scenario, but consideration needed to be given to the development from a purely navigating navigator toward a somewhat more monitoring navigator and that it would not compromise the skills of the navigator. When the IMO first started the work program of e-Navigation, there was not much evidence that, for example, large autonomous vessels would be a likely development in the foreseeable future. Today perceptions have changed and many see that it is probable that these will start to appear in the mid-2020s. The reason is, of course, because of advances in technology.

Accordingly, the IMO agreed that existing competencies and education programs should be fully used and as far as possible continue to be the core element of future mariner and other stakeholder training. New courses should be added where needed or incorporated with existing competencies and education plans, not necessarily extending the total amount of training, but rather optimizing the training to where it is most needed.

The development and use of more standardized and modularized bridge layouts may require some changes to initial training, and in the longer term there may be a potential for reducing initial and familiarization training requirements.

It was further recognized that it was equally important that all training was relevant and balanced practical skill with knowledge-based learning. Modern technology has also made possible the remote classroom through computer-based training (CBT), which might be combined with predefined classes and tutorials, free access to teaching materials, as well as online assessments and examinations [7].

Marine equipment manufacturers who have an understanding of the operational and maintenance requirements of their equipment could provide a CBT program along with the equipment to facilitate and enhance the conduct of harmonized training. In addition, CBT could also be used for onboard familiarization and refresher training, which are useful for maintaining acquired knowledge and skills.

It was repeatedly noted that lack of language skills was one of the key issues in e-Navigation user needs surveys. In order to ensure that all users had the same understanding and interpretation of communicated information, standardized phrases should be used and where needed revised or developed and their meaning published for training purposes. A similar approach was taken for the development of the IMO SMCP. Subsequently, a CBT program could be developed to help facilitate language skills.

In accordance with the strategy for the development and implementation of e-Navigation based on the identified user needs, the IMO HEAP was to be taken into account for defining the gap analysis. HEAP is a practical tool with a checklist for issues to consider, in particular relating to human element, organizational, and training issues. It was defined by the IMO in 1998 to address the human element and used for consideration of maritime safety and environmental issues.

According to the IMO's own guidance on the use of HEAP and Formal Safety Assessment (FSA) in the IMO rule-making process, HEAP "is a method developed by IMO for the use of IMO and should be seen as a practical and non-scientific checklist to assist regulators in ensuring that human element aspects related to the ship and its equipment, training, the master and the crew, management ashore and on-board and work environment conditions have been taken into consideration when introducing or amending IMO instruments" [8].

Due to considerable effort from some of the IMO member states, HEAP was systematically integrated in the final gap analysis.

At this point, preliminary shipboard user needs and priorities, preliminary shore-based user needs, and SAR authority user needs for e-Navigation were presented [9]. To illustrate this, various examples are provided. One example is about a standard interface, since mariners expressed a desire for greater standardization of functionality for navigation displays.

Navigation systems functions, operations, list of choices on the screen, and presentation (including ECDIS, radar, AIS, GPS, and GMDSS) can vary widely between manufacturers and even between models by a single manufacturer. The differences include where certain information is displayed (i.e., speed and course), how it is displayed, menu functions, and interface devices such as knobs or joysticks. This makes type-specific training difficult, and leads to ineffective use of features particularly by those watchkeepers who are new to a vessel.

Another example was bridge alert management, as bridge alerts, such as emergency alarms, alarms, warnings, and cautions should be coordinated, weighted, and support decision making without undue distraction.

It is not uncommon for the bridge of a ship to have hundreds of alarms pertaining to navigation, propulsion, cargo, and communication systems. These alarms are usually uncoordinated, physically located all over the bridge, and give little indication of severity without interrogation, which distracts the navigator.

Another example is about standardized and automated reporting to reduce the amount of ship/shore reporting and to adopt the principle of a single entry for any information into the system, including a globally standardized reporting procedure and forms to avoid repetition of reporting and to reduce workload (see Chapter 2).

A major frustration and distraction for mariners is the repeated reporting of static and dynamic information pertaining to the vessel, cargo, crew, and voyage to different shore authorities.

The identified user needs also indicated that before mariners can feel confident about relying on systems under the e-Navigation concept, they must prove to be far more reliable than any of the present systems.

Mariners sometimes struggle with electronic equipment that fails or malfunctions in some respect. This may relate to poor performance from radar, electronic chart software faults, incorrect AIS data, GMDSS alerts, or loss of position-fixing systems. Even a 99% reliability rating might result in a problem for one voyage in every 100. This has resulted in some mariners distrusting electronic systems, and now having grave doubts about relying on an e-Navigation-based system. It must be recognized that there is little competence for fixing such systems on board, and obtaining the services of a qualified technician in some ports can be difficult.

Another user need was the requirement to automatically update documents on board, such as charts and voyage planning publications, with minimal shipboard intervention.

Mariners are required to use a number of publications associated with voyage planning and monitoring. These include, but are not limited to, charts, list of lights, list of radio signals, sailing directions, and port guides. Most of these are kept on board in a paper format and require a considerable amount of time for the navigator to keep constantly updated. e-Navigation can be of benefit if it ensures that all these sources of information are automatically maintained and up to date, and all of the information is accessible from a centralized location. This may be achieved by standard formats or smart systems. The concern is that e-Navigation may lead to more information being made available, leading in turn to further overburdening. It is essential that the provision of information via e-Navigation should be managed and presented effectively.

Shore authorities have an obligation to provide maritime information to vessels. There was a need to improve the delivery and presentation of such information in order to enhance onboard decision making. Effective and harmonized communication should allow for the provision of such information in an operationally effective manner.

After discussions and approvals in the relevant IMO subcommittees, the various identified user needs were approved by the MSC.

The required functions and flows of information related to e-Navigation were identified based on the user needs. The introduction of e-Navigation required interoperability of technical components, standardization of information exchange, and automation. New functions as well as procedures needed to be integrated with existing ones. Care had to be taken that new components, roles, functions, and procedures were compatible with the existing ones. Functions of the roles related to e-Navigation were identified.

A stakeholder, such as a person or an organization, might have several different responsibilities or roles. An example is the captain's formal responsibilities.

The captain is formally responsible for all functions carried out onboard. Although the captain delegates tasks to crew members such as the bridge team, the formal responsibility remains with him or her. Information processing functions conducted on the ship's navigation bridge were divided into the following subfunctions:

- Safe navigation;
- Support and control of safe navigation;
- Management of information;
- Support of incident handling and emergency management;
- Support maritime security.

Several conventions, regulations, and practices define the rules and premises for these functions (e.g., SOLAS, STCW, and ISM code.

Ship- and shore-based users have a wide range of identified responsibilities connected with the safe berth-to-berth navigation of ships. Supporting these sets of responsibilities are roles that require functions to be carried out or activities to be executed that in turn contribute to the fulfillment of those responsibilities.

Within e-Navigation, some functions may be automated, while others will continue to be carried out though interactions between persons and systems. Seamless information flows and improved access to electronic information will enable automated assessments across different information sources. Information should be useful by being provided to best aid decision support by increasing situational awareness. This should lead decision makers to consistently make appropriate decisions and to take reliable and effective actions.

Several roles may be involved in a process, and there are interactions between functions, as given in the previous example of a captain's formal responsibilities, belonging to a role or different roles.

Interactions between the functions identified in the process descriptions were defined by means of information elements. Information elements were combined into specifications of the content of the information that should be exchanged, such as information exchange between onboard and ashore systems. Consequently, the process description and the information exchange interfaces were together with the user needs as the basis for the specification of electronic services provided from ashore to ships.

This also represented an important outline of the development of functional architecture that assisted in achieving the objective of e-Navigation. The architecture safeguarded that work is conducted in a systematic manner and that processing of information continuously was kept at the center of the work.

3.2 Gap Analysis

The next step was to perform a gap analysis, where user needs formed an important starting point. The follow-up gap analysis focused on main elements such as operational, technical, and regulatory gaps in addition to a training need analysis based on the system architecture and operational concept.

According to IMO, the gap analysis should focus on the following elements:

1. Regulatory gap analysis particularly focused on identifying gaps in the present frameworks that needed to be filled;
2. Operational gap analysis to define a reduced concept of operations that could be used based on the integration of existing technology and systems;
3. Identification and description of existing systems that could be integrated into the e-Navigation concept covering functionality, reliability, operational management responsibilities, regulatory status as to specification/standardization, fitment and use, and generational status and integration with e-Navigation system requirements;
4. Technical gap analysis comparing the capabilities and properties of existing systems with architectural requirements to identify any technology or system development that might be needed, based solely on user needs [10].

A gap analysis determines the space between where we are and where we want to be, and serves as a means to bridge that space. The reason for performing the e-Navigation gap analysis was to use it as a tool for identification, filtering and mapping of agreed-on user needs, identifying gaps, and finding solutions. User needs from the shipboard, shore-based and SAR fields might already be fulfilled by current maritime functions and services. If not, there were certain discrepancies between what users require and the current situation. That turned out to be a rather challenging work, as an increasing number of factors with influence and relevance to the process were identified. After a series of preliminary proposals, the representatives of the ROK provided a comprehensive table [11] to the IMO taking into account all relevant factors and information. It took the IMO some time to get familiarized with and to digest this table because of its complexity, but it ended up as the basis for the final presentation of identified gaps and proposed practical e-Navigation solutions to address those gaps. It was considered sufficiently progressed to be applicable to the further process of identifying risk reduction and cost benefits of the e-Navigation solutions.

Gaps derived from identified user needs were therefore reviewed and identified as operational, technical, regulatory, and training. These gaps for shipboard, shore-based, and SAR users were further categorized as follows:

- Information/Data Management
  - Common data structure/harmonized data formats
- Effective and Robust Voice Communication and Data Transfer
- Navigational Bridge Systems and Equipment
  - Improved reliability and indication of reliability
  - Improved ergonomics, standardization, and alert management
  - Presentation of information received via communication equipment (e.g., MSI) on the navigation display
  - Documents in electronic form and automated updates of information
- Ship Reporting
- Training and Familiarization
- Traffic monitoring

More details about gaps are discussed in Section 3.3.

For each addressed gap there were proposed practical e-Navigation solutions applying to the four categories. Each piece of information about a gap was also linked to its origin. This included category and subcategory, user needs, key strategic elements, core strategic objectives, related functions, existing equipment, and operational area.

The IMO agreed that the implementation of e-Navigation should be a phased iterative process of continuous development, including the process shown in Figure 3.2.

At the same time, the IMO invited the IHO to report to them and provide updates on an annual basis regarding progress made in worldwide ENC coverage. As listed below, ENC is one of the key strategy elements of e-Navigation [12].

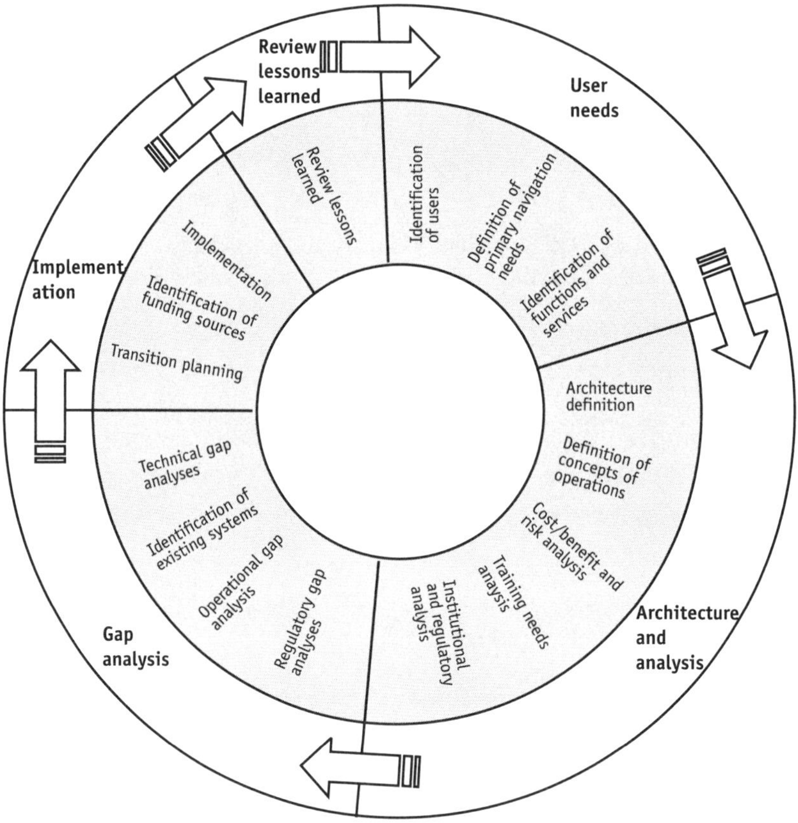

Figure 3.2 The e-Navigation wheel, which shows the components of the development of the e-Navigation implementation process. (IMO document NAV54/25, Annex 12.)

Identified gaps and proposed solutions were related to shipboard, shore-based, or SAR users, but also to the various categories of the gap analysis such as the main groups of gaps listed above (see Figure 3.3). Accordingly, seven subcategories of the gap analysis were listed.

The gaps were then described with reference to defined user needs, agreed-on key elements of e-Navigation, the core objectives of the e-Navigation strategy, related functions, existing equipment, systems and technologies, and actual operational area.

From the outset, eight key strategy elements based on user needs were listed:

1. Architecture;
2. Human element;
3. Conventions and standards;
4. Position fixing;
5. Communication technology and information systems;
6. ENCs;
7. Equipment standardization;
8. Scalability.

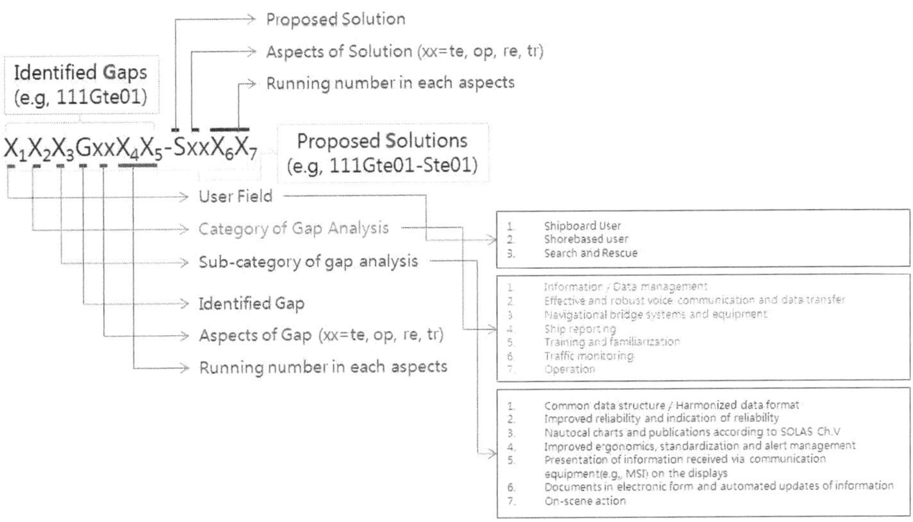

Figure 3.3 Running number for gap analysis. (IMO document COMSAR 16/11.)

The overall conceptual, functional, and technical architecture, which is described in greater detail in Chapter 5, should be developed and maintained, particularly in terms of process description, data structures, information systems, communication technology, and regulations.

In general, knowledge of a particular domain is an integral part of working competently, effectively, and successfully within that domain. The focus was on the rationale for the development of instruments to measure the information system's knowledge of individuals.

In this respect, training, competency, language skills, workload, and motivation were identified as essential. Alert management, information overload, and ergonomics were prominent concerns. These aspects of e-Navigation were taken into account in accordance with IMO's human element work.

Furthermore, position-fixing systems should meet user needs in terms of accuracy, integrity, reliability, and system redundancy in accordance with the level of risk and volume of traffic.

It was important that communications and technology information systems should be identified to meet user needs. This work might involve enhancement of existing systems or development of new ones. Any impacts affecting existing systems would need to be identified and addressed, based on technical standards and protocols for data structure, technology, and bandwidth and frequency allocations.

An important task for maritime safety and security, which is one of the core objectives of e-Navigation, is anomaly detection. A situation awareness system must support navigators and VTSOs in identifying and focusing on suspicious or abnormal situations while filtering out normal situations as much as possible. A system supporting situation awareness must present its output in a useful and manageable way to a human user. This was taken into account during the development of Guidelines on Software Quality Assurance and Human-Centred Design for e-Navigation, as described in Chapter 1. The guideline also supports the principles identified in SOLAS Convention regulation V/15 relating to bridge design, design and arrangement of navigational systems and equipment, and bridge procedures.

3.3 Solutions Identified by the Gap Analysis

The proposed practical e-Navigation solutions to address identified gaps were listed with reference to their operational, technical, regulatory, and training relevance.

Addressing the gaps related to navigational bridge systems and equipment, and with special reference to improved ergonomics, standardization, and alert management, it was underscored that the ergonomic problems of navigation equipment exist in a sense that there is a lack of intuitive HMI for communication and navigation means. Bridge layouts, equipment, and systems are seldom designed from an ergonomic and/or user-friendly perspective and control (e.g., type approval) of software and hardware updates is not sufficient. Special attention was paid to the presentation of information received via communication equipment (e.g., MSI) on the navigation display. In this respect, a test bed was carried out in Singapore.

In 2012, the Norwegian Coastal Administration (NCA), assisted by the Maritime and Port Authority of Singapore (MPA) and IMO's MEH project in the Straits of Malacca and Singapore, hosted a workshop in Singapore. The aim of the workshop was to demonstrate the use of the IHO Universal Hydrographic Data model, S-100, and consider potential synergies between e-Navigation and the MEH project in the Straits of Malacca and Singapore.

The test bed demonstrated the use of IHO S-100 to promulgate MSI using AIS and Notice to Mariners (T & P notices) and Wimax to access the Internet. The demonstration was to verify the concept of communication between shore and ship in transmission of safety and environmental information through the use of AIS and Internet access at sea.

The scenario chosen was a collision involving two vessels, with one of them potentially sinking. In response, the VTSO generated a virtual (AIS) aid to navigation and MSI. The MSI was transmitted via the AIS VHF data link. It was successfully received on board, converted to S-100 XML format, and displayed on the bridge. A report on the outcome of this test bed was submitted to the IMO [13].

While the IMO was developing potential e-Navigation solutions, it discussed effective and robust voice communication and data transfer. Insufficient reliability of data/voice communications (users require communication without interference, disruption, and noise), the lack of reliability standards for communication technology and the possible lack of bandwidth and unclear assignment of adequate bandwidth for potential e-Navigation communication needs were listed as relevant challenges.

For shore-based users, IMO noted shore-to-ship bandwidth limitations, shortage of VHF marine frequencies, and a lack of effective means to reduce the transmission of inaccurate AIS data.

And finally, for SAR users, a lack of mechanisms to provide SAR rescue coordination center (RCC) functions with a full range of relevant e-Navigation information in digital format was addressed.

On the shipboard side, it was noted that there are no standardized data formats established for ship reporting. Further, there is a lack of technical means to make information of a vessel's intentions available to relevant user stakeholders and a lack of assessment procedures to quantify reliability parameters such as specific assessment of electronic position-fixing systems.

For improved ergonomics and standardization of alert management, a lack of standardized symbology of all information required to display on the navigational system was noted, as well as insufficient reliability of position-fixing systems, a lack of self-checking functionality of the electronic equipment for improved reliability, a lack of automatic assessment functionalities to provide quantified reliability information, a lack of position, navigation, and timing (PNT) relevant services for port operation and automatic docking, a lack of framework for provision of resilient PNT, a lack in presentation of maneuvering information/data (engine room telegraphs) on navigational display, and a lack of standardization for operation of functions to observe the passage plan. Users require standardization on the level of function provided and the functional way of operating it, but not being restricted to future developments.

For the presentation of information received via communication equipment on the navigational display, there was a lack of user-selectable and task-oriented presentation, unavailability of information in real time with possible presentation on the navigational display to support bridge operation, lack of integrated secondary screen option for digital publications and MSI, and a lack of information about special berthing requirements on navigation systems, especially for piloting.

Transnational reporting requirements are not harmonized, and there is a lack of standardized reporting formats. Reporting procedures should be globally standardized.

As for training, there is insufficient familiarization material for safety-related equipment, and a lack of specifications of familiarization material for new and existing performance standards.

The gap analysis also showed that for shore-based users, there was a lack of common maritime information/data structure harmonizing the policies for the security and use of data and insufficient identification of harmonization needs for standards, formats, and protocols. Furthermore, it also identified lack of protocols, formats, and data structure that enable shore-based authorities to exchange information with other authorized shore-based users. There was a gap between information capability of current information management systems and those that will be required as volumes of information increases, and tools that have the capability to manage increased levels/volumes of information are not in use.

There are inconsistent rules that require some coastal states to maintain domain awareness, insufficient collection of data required to establish accurate and reliable marine domain awareness, and a lack of international guidance on security of data and its sharing.

Addressing systems and equipment, the IMO noted that there is insufficient delivery and presentation of maritime information that shore-based authorities are required to provide to ships. There were no standard data formats for onboard capture and presentation that cover the entire scope of information provided by VTS, and there was a lack of harmonized presentation of domain awareness to improve situational awareness for allied and other support services.

For traffic monitoring, it was noted that the tools that have the capability to manage increased levels/volumes of information are not in use. Current VTS infrastructure may not have the capacity for increased collection, integration, exchange, presentation, storage, and analysis of data, and there is a lack of procedures that enables shore-based authorities to monitor quality of navigation systems on board as well as quality of information and effectiveness of communication.

During the consideration of an e-Navigation solution concerning automated and timely updating of ENCs, nautical publications, and other documentation, some delegations emphasized that the use of paper charts should be taken into account as it still existed on board many ships.

Furthermore, the IMO Formal Safety Assessment (FSA) noted the complexity of the subsequent risk and cost benefit analysis process, and decided to group potential e-Navigation solutions into the following broad categories to facilitate completion of the analysis [14]:

- Improved, harmonized and user-friendly bridge design;
- Means for standardized automated reporting;
- Improved reliability, resilience, and integrity of bridge equipment and navigation information;
- Integration and presentation of available information in graphical displays received via communication equipment;
- Information management;
- Improved access to relevant information for search and rescue;
- Improved reliability, resilience, and integrity of bridge equipment and navigation information for shore-based users;
- Improved and harmonized shore-based systems and services;
- Improved communication of VTS service portfolio.

After finalizing the gap analysis, IMO proposed a preliminary list of potential e-Navigation solutions, and further proposed that this list should be used as the basis for further identification of risk control options as preparation for the subsequent risk and cost benefit analysis, which was the IMO FSA, the methodology used in the IMO rule-making process.

This concluded the identification of gaps based on registered user needs, and thereby defining eventual risk reductions and cost beneficial solutions to be further processed. The effort of carrying through a very complex analysis of gaps influenced by a great number of factors seemed to be worthwhile as the end result was rather clearly defined solutions. Of nine original e-Navigation solutions, five of them were prioritized as described in Chapter 1.

The process had been extensive with annual meetings at IMO at the MSC committee, and the subcommittees NAV, COMSAR, and STW, and at later stages annually at the MSC committee and the subcommittee NCSR, in addition to intersessional work of the correspondence group on e-Navigation for a five-year period. It involved the participation and cooperation of a great number of IMO member states and international organizations. This included, among others, IHO and IALA, who were at the same time active participants in the e-Navigation efforts at IMO and in parallel worked on specific e-Navigation issues within their areas of responsibility.

3.4 e-Navigation Development by IHO and IALA

3.4.1 IHO

Hydrographic services include, according to SOLAS, "the collection and compilation of hydrographic data and the publication, dissemination and keeping up to date of all nautical information necessary for safe navigation" [15]. The provision of "all nautical information necessary for safe navigation" in digital format is a key-enabler for e-Navigation [15]. That objective is partially met with the progressive uptake of ECDIS on ship bridges and the provision by hydrographic offices of the relevant integrated charting services to ensure that mariners have access to up-to-date seamless digital chart coverage of ENCs through the IHO Worldwide ENC Database.

The IHO is the intergovernmental organization responsible for creating a global environment in which the hydrographic offices of its member states, currently numbering 85, provide collectively adequate, standardized, and timely hydrographic services and ensure their widest possible use.

The implementation of e-Navigation requires extending the ECDIS concept from chart information to include additional relevant nautical information. Driven by the constraints and limitations associated with traditional paper-based products, the carriage requirements set out by SOLAS differentiate between nautical charts and nautical publications, such as sailing directions, lists of lights, notices to mariners, tide tables, and all other nautical publications. This distinction is no longer relevant when the information is managed and processed digitally through electronic display and information systems. The need to develop solutions that support mariners with an integrated real-time situational awareness is recognized in the SIP. It relates directly to one of the five prioritized e-Navigation solutions on which the SIP is based, namely solution S4: "integration and presentation of available information in graphical displays received via communication equipment" [15]. As indicated in the SIP, the implementation of solution S4 requires the development of the concept of MSPs, which is the core of solution 5 and define and describe the sets of operational and technical services and their level of service provided by a stakeholder in a given sea area, waterway, or port, as appropriate. The implementation of MSPs will be supported by a common maritime data structure (CMDS) based on the S-100 Universal Hydrographic Data Model. CMDS is also important for the development of the e-Navigation solution for automated ship reporting.

S-100 has been developed by the IHO since 2001 as a successor to IHO Standard S-57, Transfer Standard for Digital Hydrographic Data. S-100 aims at providing a contemporary hydrographic geospatial data standard aligned with the current ISO 19100 series of geospatial standards. S-100 specifies, for hydrographic and related information, the methods and tools for data management, processing, analysing, accessing, presenting, and transferring such data in digital/electronic form between different users, systems, and locations. A major advantage of S-100-based products is the use of a fully machine-readable catalog system that will allow the ECDIS to update new elements via a catalog update, which comes together with a regular data delivery.

The first edition of S-100 was published by the IHO in January 2010. A revised and enhanced second edition was published in June 2012. It is expected that further evolution will be required to support streaming data services related to e-Navigation.

The S-100-based MSP related to hydrographic services will be structured in several products and services whose features, attributes, and relationships are described in S-xxx product specifications. A new generation of ENCs, defined by the S-101 Product Specification, will be used as the basis for the mariner to combine, as required, in an integrated display with additional layers related to a more detailed description or to other products or MSPs. Table 3.1 lists the product specifications identified by IHO that need to be developed.

The development of S-101 reflects the experience and feedback of end users and other stakeholders in regard to the current limitations of S-57-based ENCs. The ultimate goal is that an S-101 ENC will provide the base chart layer for S-100 integrated products and the underpinning of e-Navigation. Its full potential will be realized as other types of products used within the maritime domain, such as aids to navigation, vessel traffic services, oceanography and meteorology, adopt the concept and develop interoperable products and services also modeled along the S-100. It is expected that the navigation equipment industry will develop S-100-ready devices when the S-101 Product Specification reaches maturity and regular provision of S-101 comes within sight.

Ensuring the harmonized display of information related to more than one navigational data product is part of the broader

Table 3.1
IHO Product Specifications Based on S-100

Product Specifications of IHO (Numbers S-101 to 199)	
S-101	Electronic Navigational Chart (ENC)
S-102	Bathymetric Surface
S-103	Subsurface Navigation
S-104	Tidal Information for Surface Navigation
S-111	Surface Currents
S-112	Dynamic Water Level Data
S-121	Maritime Limits and Boundaries
S-122	Marine Protected Areas
S-123	Radio Services
S-124	Navigational Warnings
S-125	Navigational Services
S-126	Physical Environment
S-127	Traffic Management
S-128	Catalogues of Nautical Products

Source: IHO 2016.

issue of interoperability between the relevant product specifications. Beyond the consistency ensured through compliance with the S-100 data model and the S-100 Registry structure, it is necessary to develop an S-100 Interoperability Specification that considers the following:

- "Harmonization of portrayal between different data products, e.g., to ensure that less significant features in one data product, are not displayed more prominently than more significant features in another product" [16];
- Determining what features in the base S-101 ENC should be turned off when the operator turns on different overlays such as sea ice and weather;
- Determining different display priorities between multiple S-100-based products;
- Testing procedures to ensure that data products are not only tested against their own performance criteria, but also tested against their intended interaction with other data products [16].

An overarching test bed strategy was developed in 2014 to address the whole logical progression from individual feature and portrayal catalog creation to the use of products within an ECDIS through intermediate steps, such as the display of products in simple viewers and the validation of production tools.

Further information on S-100 and related developments are available on the IHO website (www.iho.int; see the section on ENCs, ECDIS, and S-100). The information includes sections such as "Master Plan for the Development and Implementation of S-100" and "S-101 Value Added Roadmap," which provide an estimated sequence of events for the development, testing, and implementation of S-101.

The agreed-on IMO-IHO harmonization work on data modeling decided by IMO (MSC 90) focuses on the relevant outputs related to the IMO e-Navigation SIP. The work includes the need for "some form of overarching coordination to ensure the ongoing management and maintenance of the (maritime information and data) structure" [17] and the group was tasked to "consider matters related to the framework for data access and information services under the scope of SOLAS" [17].

3.4.2 IALA

The purpose of the IALA is to ensure that seafarers are provided with effective and harmonized marine AtoN services worldwide to assist in safe navigation of shipping and protection of the environment. This also includes VTSs and communications. The IALA organizes e-Navigation conferences together with national authorities and other stakeholders and gathers together marine aids to navigation authorities, manufacturers, consultants, and scientific and training institutes from all over the world and offers them the opportunity to exchange and compare their experiences and achievements. Taking into account the needs of mariners, developments in technology, and the requirements and constraints of aids to navigation authorities, a number of technical committees have been established, bringing together experts from around the world.

The work of the committees is aimed at developing common best practice standards through publication of IALA standards, recommendations, and guidelines.

This work ensures that mariners have AtoN that will meet their needs both now and in the future. The IALA has encouraged cooperation between nations to assist developing nations in establishing AtoN networks in accordance with the degree of risk for the waterway concerned.

The implementation of e-Navigation connects the ship and the shore digitally in a number of ways and for a number of reasons. Solution 5 of the e-Navigation concept (MSPs) makes sure that shore-based information received on board from ports, VTSs, and other service providers to ships is harmonized and standardized around the world.

Furthermore, the IALA reviews S-201, which would be the first IALA data standard on AtoN information based on the IHO S-100 data series of standards, and the organization has also been looking at management of the IALA domain under the IHO registry of domains. It has subsequently prepared a guideline on the subject. The IALA has, in addition, recognized the need for S-100 to support streaming data.

Regarding implementation, the association has been responsible for gathering information on all e-Navigation test beds. In this regard, the IALA guidelines on test bed reporting has been revised and published in June 2016.

On telecommunications, the IALA has been working on a new technology called VDES. This technology utilizes a part of the VHF spectrum in the maritime mobile band to transmit and receive data both terrestrially (VDE-TER) and with satellites (VDE-SAT). Work is continuing on the message structure for such services and preparing IALA recommendations as well as inputs to the ITU, supporting their work on frequency allocation for VDES. Important work has been carried out on e-Navigation services and the IALA continues to work on the guidelines for MSPs. It is clear that there are a number of stakeholders for portfolio services who are involved in the drafting of the guidelines. The IALA is clearly responsible for the first three MSPs, MSP1-VTS Information Service, MSP2-Navigation Assistance Service, and MSP3-Traffic Organization Service. For each MSP there will be a number of detailed attributes.

When it comes to PNT, the task has been to ensure that traditional and new navigation systems are available for e-Navigation. One of the key issues has been a terrestrial backup and augmentation to the existing GNSS system. This includes work on new concepts such as e-Loran and R-Mode, and utilization of signals of opportunity for PNT.

The IALA has been aware that some new systems may have patent issues and it has developed a policy to ensure that for any new system being considered for e-Navigation that may potentially infringe a patent, the requirements of the patent are declared before adoption. This issue caused a lot of legal problems at the introduction of AIS.

The IALA is committed to the development of e-Navigation and is able to provide a good forum for this work; however, it is somewhat constrained by its constitution to shore-based services only. e-Navigation crosses this boundary.

With a vision of safe and clean seas and saving lives, and a mission to ensure safe vessel operations, combat marine pollution, and rescue people in distress, many IMO member states and international organizations have stated that they will continue to contribute to the development of e-Navigation in the areas where they have expertise and can make a meaningful difference. Chapter 4 shows some examples of this work.

References

[1] Maritime Safety Committee, IMO, Development of an E-Navigation Strategy, Document MSC 81/23/10, December 19, 2005 [online], available at https://www.iho.int/mtg_docs/com_wg/HSSC/CHRIS18/CHRIS18-04.4C_e-Navigation_strategy.pdf, accessed May 10, 2017.

[2] Wadsworth, B., Marine eNavigation: An Orientation Paper, March10, 2005, available at https://www.iho.int/mtg_docs/circular_letters/english/2005/Cl29e.pdf, accessed May 10, 2017.

[3] IMO, Development of an E-Navigation Strategy Implementation Plan, Document NAV 56/WP.5/Rev.1, July 28, 2010 [online], available at https://www.iho.int/mtg_docs/com_wg/CPRNW/S100_NWG/2014/NAV%2056_WP5_rev1.pdf, accessed May 10, 2017.

[4] IMO, Document NAV 55/WP.5, Annex 2.

[5] IMO document HTW 42/6.

[6] IMO, Development of an e-Navigation Strategy Implementation Plan, Document COMSAR 15/11, December 3, 2010 [online], available at https://www.iho.int/mtg_docs/com_wg/CPRNW/S100_NWG/2014/COMSAR%2015-11_e-Nav_CG.pdf, accessed May 10, 2017.

[7] IMO, Report to the Maritime Safety Committee, Document STW 43/14, June 11, 2012 [online], available at https://www.uscg.mil/imo/htw/docs/stw43-report.pdf, accessedMay10, 2017.

[8] IMO, Guidance on the Use of Human Element Analysing Process (HEAP) and Formal Safety Assessment (FSA) in the IMO Rule Making Process, Document MSC/Circ.1022, MEPC/Circ.391, May 16, 2002 [online], available at http://www.safedor.org/resources/1022-MEPC391.pdf, accessed May 10, 2017.

[9] IMO, Document NAV 56/8, Annex 2, 3, and 4.

[10] IMO, Document MSC 85/26/Add.1 ANNEX 20 [online], available at http://www.imo.org/en/OurWork/Safety/Navigation/Documents/enavigation/MSC%2085%20-%20annex%2020%20-%20Strategy%20for%20the%20development%20and%20implementation%20of%20e-nav.pdf, accessed May10, 2017.

[11] IMO, Document COMSAR 16/11.

[12] IMO, Document NAV 54/25 Annex 12.

[13] IMO, Document NAV 58/INF.4.

[14] IMO, Document NAV 58/WP.6 Rev1.

[15] IMO, Implementation of Solas Regulation V/9. Hydrographic Services, Document MSC/Circ.1118, May 27, 2004 [online], available at http://solasv.mcga.gov.uk/msc/MSC-Circ%201118.pdf, accessed May 10, 2017.

[16] IHO, Interoperability of S-100 Product Specifications, Document S100WG-10.2A [online], available at http://www.iho.int/mtg_docs/com_wg/S-100WG/S-

100WG1/S100WG01-10.2A%20%20Interoperability%20Specification.pdf, accessed May 10, 2017.

[17] IHB, Comments on paper HSSC8-07.2A–IMO Activities Affecting HSSC (Including E-Navigation), Document HSSC8-07.2B, October 24, 2016 [online], available at https://www.iho.int/mtg_docs/com_wg/HSSC/HSSC8/HSSC8-07-2B_Comments-on-HSSC8-07-2A.pdf, accessed May 10, 2017.

4

e-Navigation Solutions

4.1 Introduction to e-Navigation Solutions

As discussed in Chapter 3, while the first steps to a cohesive e-Navigation plan involve implementing the five prioritized e-Navigation solutions, it is important to recognize that further e-Navigation development will be a continuous process following user needs for additional functionalities of existing and possible future services, systems, and equipment. As user needs evolve and new technology is introduced, other e-Navigation solutions might be incorporated. However, the main objective of the IMO's e-Navigation SIP is to implement the five prioritized e-Navigation solutions. In order to do so, a number of tasks have been identified by several international organizations and many projects and test beds have been established. Some countries have already started the implementation of e-Navigation and there is ongoing work that will provide the industry with harmonized information and new standards regarding the design of products and services to meet the e-Navigation solutions.

4.2 Further Development

It was realized that further development could only take place if supported by projects that incorporated some sort of demonstration and testing arrangements known as test beds, which support the approved e-Navigation solutions. There are many such projects running that are clarifying the performance of e-Navigation. Some of the test beds are described below in relation to each solution.

In July 2016, the government of the Russian Federation approved an innovative project for creation of an e-Navigation test bed within its strategic National Technologic Initiative government program.

The main project participants were the Federal Marine and River Transport Agency, Federal State Unitary Enterprise Rosmorport, and Transas Marine Ltd. The following goals of the project were declared:

- Developing and testing e-Navigation solutions based on the existing infrastructure of regional VTS in the Gulf of Finland and participating vessels;
- Implementation of developed solutions and services available for use by different stakeholders (VTS, vessels, SAR coordination centers, port administration, ship owners and agents, tug operators, pilot organizations, etc.);
- Evaluation of implemented solutions and services for possible future implementation in other coastal areas within Russian territorial waters and approaches.

4.2.1 Solution 1: Harmonization of Bridge Design

Work has been carried out on relevant IEC standards supporting the harmonization of bridge design and display of information. Work has also been undertaken by Australia, the ROK, and some international organizations for the development of guidance on the S-Mode of operation of navigation equipment, including the interrelation between the INS and S-Mode.

In support of solution 1, it has been underscored that there should be a functionality to distinguish navigation safety-related information from ordinary business information, including issues related to gateways, two-way communication between navigation and communication equipment, store and forward capability, and bridge alert management.

Parallel work has been carried out on the other e-Navigation solutions. In addition, several test beds and projects have been established. The outcome of these test beds and projects has led to an understanding of how the e-Navigation solutions best could be put into effect.

4.2.2 Solution 2: Means for Standardized and Automatic Ship Reporting

Since 2014, China has been conducting studies on standardized and harmonized ship reporting systems, which is in line with solution 2.

In 2015, China developed and demonstrated a unified electronic ship reporting system in Shenzhen Port. This system supports the data structures and formats defined by the IMO. It provided unified and standardized electronic ship reporting service to shore-based users and participating ships.

This confirms the need for electronic ship reporting systems in general and the advantages of providing the information to shore-based stakeholders; however, solution 2 has an important role for the ship to reduce workload by preparing information automatically for transmission from the ship.

In 2015, the IMO welcomed a proposal from Norway, Brazil, and Singapore to conduct a test bed to prove technology for ship reporting and which also supported the revision of Guidelines and Criteria for Ship Reporting Systems. The use of the test bed provided a better understanding of how to revise the guidelines [1].

The test bed setup was the installation of a test unit on a ship and installation of a system on shore for receiving and integration into the Norwegian Maritime Single Window (Norwegian SafeSeaNet), which is described in more detail in Section 4.3.3. The data transmitted was a subset of the single window reporting data set, such as arrival information, ship security information, and border control.

AIS and application-specific messages (ASMs) were used for the ship- and shore-based communications solutions. Norwegian SafeSeaNet was used as the final destination. IMO and IEC technical standards for AIS and VHF were complied with.

The assessment of the test bed was that it was a successful trial of the concept of electronic transfer of data. Further development and testing of other communications systems such as VDES and improved cellular technology is required in order for a complete harmonized system to be adopted.

This test bed showed that electronic ship reporting systems and electronic prearrival information, as described in the approved e-Navigation solution 2, worked satisfactorily. The results also showed that the technology was available for automated ship reporting.

4.2.3 Solution 3: Improved Reliability, Resilience, and Integrity of Bridge Equipment and Navigation Information

The EU project ACCSEAS was presented in Chapter 1 and was a three-year project supporting improved maritime access to the European North Sea region through minimizing navigational risk. With European transport policy providing a shift to seaborne transport, using short sea shipping to avoid road bottlenecks to the movement of goods, services, and people, efficient and effective marine navigation services have never been more important.

A test bed carried out in the project was to pilot reliable ship positioning with associated shore infrastructure and a vessel integrated navigation system based on the e-Navigation architecture. One of the services designed in the test bed was resilient position navigation timing (resilient PNT). This test bed supported e-Navigation solution 3.

On several excursions aboard the Trinity House Vessel (THV) Galatea out of Harwich, UK, the ACCSEAS project described in Chapter 1 successfully demonstrated a prototype resilient PNT system. The system used an alternative technology, e-Loran, to automatically and seamlessly step in to provide PNT in the event of GPS loss or failure [2].

4.2.4 Solution 4: Integration and Presentation of Available Information in Graphical Displays Received via Communication Equipment

Another important project was the German e-Navigation test bed TICON, putting the IMO's e-Navigation solution 4 to the test.

"The current separation of communication and navigation systems does not support permanent availability and integration of the latest safety-related and geospatial information for the situational awareness and decision-making required for safe navigation" [3].

The test bed focused on "integration of communicated geodata and/or safety-relevant information (water level, current, ice, weather, MSI) into onboard navigational systems" [3] by utilizing a shore-based data clearing house concept.

Examples of exchanged information were MSI, improved depth information, wind and current information, ice information, and SAR needs.

In port, a ship received basic data sets and updates prior to departure via shore-based communication or external storage device, called large data volume information, which included static data such as ENCs, which was integrated into the onboard database and made available in a user-friendly manner for INS task displays (route planning) [3].

The "underway ship receive[d] updated data and forecasts via wide-area (e.g., 3G, satellite) communication systems, called small data volume," [3], tailored to the ship's need to reduce data volume. The tailoring took into account ship-specific data, such as draught and voyage plan [2].

4.2.5 Solution 5: MSPs

As part of the improved provision of services to vessels through e-Navigation solution 5—improved communication of VTS service portfolio, not limited to VTS stations—the core of this solution, MSPs have been identified as the means of providing electronic information in a harmonized way. The relevant services, as currently defined by the SOLAS Convention, cover a broad scope, including aids to navigation, hydrographic services, maritime safety information, meteorological services, pilotage, and vessel traffic services. The list of agreed-on MSPs are seen in Table 4.1.

A lack of coordination in the provision of information related to maritime services and among organizations responsible for the provision of MSPs may lead to duplication of efforts, development

Table 4.1
MSPs

MSP1	VTS information service
MSP2	Navigational assistance service
MSP3	Traffic organization service
MSP4	Local port service
MSP5	Maritime safety information service
MSP6	Pilotage service
MSP7	Tugs service
MSP8	Vessel shore reporting
MSP9	Telemedical maritime assistance service
MSP10	Maritime assistance service
MSP11	Nautical chart service
MSP12	Nautical publications service
MSP13	Ice navigation service
MSP14	Meteorological information service
MSP15	Real-time hydrographic and environmental information services
MSP16	Search and rescue service

of regional solutions, use of different communication systems, and the provision of superfluous or noninteroperable information.

The technical development of MSPs has been about harmonizing the format, structure, and communication channels for the electronic exchange of information between shore and ship, and is coordinated by the IMO. However, the content of MSPs are being developed by different international organizations. Therefore, coordination among the organizations ensures harmonization of scope, format, structure, display on board, and communication systems used to transmit the information electronically.

IMO states, "MSPs containing georeferenced information can be displayed on equipment such as ECDIS and radar. The interoperability of the MSPs must also be ensured to provide mariners with integrated real-time situational awareness" [4].

Next are some examples of test beds and work on MSPs that have been carried out.

As described in Chapter 1, the ship traffic management system provided from Singapore VTS optimizes sea traffic flow and provides vessels with advice on the best speed and route. This enables a vessel to arrive at a port at the optimal time or create an optimal transit passage. In this way, it is possible to avoid traffic congestion, reduce fuel consumption, reduce emissions, and increase overall safety and efficiency. This will enable just-in-time arrival of vessels, improve the efficiency of existing infrastructure, and reduce the environmental footprint [5].

The core of the concept was that ship systems on board, the ship traffic management system at the Singapore VTS and the ship traffic simulator (STS) can share information with each other regarding route optimization and arrival time according to the principles of just-in-time arrival. With this interaction, the ship traffic management system can receive a route from the ship's ECDIS and send the route to the STS. Once the STS has performed its calculations and optimized the route, the result can be displayed on the ship traffic management system, which can then forward the proposed optimized route and speed to the ECDIS. This concept supports MSP 3 traffic organization service (TOS).

Another project worth mentioning took place in the harbor of Tianjin in North China that established a major demonstration project called e-Navigation. The aim was "to provide timely, complete, and comprehensive maritime safety and navigational aids [through building a standard basic system] which will help to improve transportation efficiency, guarantee navigation safety, and enhance environmental protection" [6].

In Yangshan Port of Shanghai, a marine meteorological observation system is installed based on buoys, which support MSPs 14 and 15—meteorological information service and real-time hydrographic and environmental information services. Data collection is automatic.

"The basic composition of a marine meteorological buoy includes a buoy, mooring system, sensor, data acquisition, and processing system, communication system, power system, detector, and shore-based receiving and processing systems" [7].

Over the past century, health care on land has gone through tremendous changes. Technological and scientific progress has greatly helped medical practitioners improve patient treatment. Nevertheless, little has changed with regard to medical care for people working at sea. Natalia Tunheim was a student at the Oslo School of Architecture and Design in Norway, and in her diploma project "Designing for Health Care at Sea," she explored the different interconnected systems put into play when an accident or illness occurred on board vessels. The aim was to identify opportunities to improve information flow when collaboration between medical practitioners on land and at sea is necessary. This work also supported MSP 9—telemedical maritime assistance service. She mapped out the landscape before, during, and after an acute injury or illness occurs.

Among her findings, she concluded that labeling of medical equipment by prototyping a set of symbols, constraints, and opportunities in a video link on board the vessel and utilized unused and unexploited resources by taking advantage of the time that was already set aside for internal exercises in medical care, helping the crew toward purposeful practice. The final delivery was EASYMED—a medical treatment course that was carried out on board the vessel.

The subject plan centered on learning objectives with regard to the medical needs and responsibilities on board the vessel and of shore-based emergency service, both of which have to interact well together [8].

In support of MSP 16—search and rescue (SAR) service, the primary aim of a test bed carried out in Chile was to demonstrate the possible use of the existing LRIT shore-based infrastructure to exchange additional ship information between authorities, other than LRIT information, in accordance with the prioritized e-Navigation solutions, and to facilitate, in particular, SAR operations in remote and polar areas.

The test bed used a SAR scenario off the Chilean coast, using LRIT to retrieve a SAR surface picture of two predefined ships within a specified area.

An "add more information" option was integrated to retrieve more detailed information about the ship sailing from Norway to Chile from the ship reporting system SafeSeaNet Norway.

Additional information consisted of a picture of the ship, port of destination, number of persons on board, dangerous goods carried on board, and latest satellite AIS position.

This information was displayed in less than 3 seconds after sending the request from the Chile National Data Center (NDC) to the Norwegian Single Window. Below is a list of advantages to this process:

- During the demonstration, e-Navigation information was retrieved upon automatic request of the Chilean NDC to the Norwegian Single Window;
- The systems were interconnected via web services and operated live;
- Relying on the S-100 Universal Hydrographic Data Model, new attributes were easily added using the existing format of messages exchanged (e.g., number of crew and persons on board, last port of call, type of ship, cargo).

Another project that has demonstrated MSPs was the EU-financed Sea Traffic Management (STM), led by Sweden, which intended to overcome many of the challenges of communication and information sharing between stakeholders in the maritime transport industry. By providing vessels with the ability to see each other's planned routes, navigators get a more complete picture of how surrounding vessels will influence their onward voyage. Using this data, other services are able to produce valuable information and offer advice to vessels on their routes, such as recommendations to avoid congestion in areas with high traffic, avoidance of environmentally sensitive areas, and maritime safety information. The information exchange between vessel and port actors will improve planning and performance regarding arrivals, departures, and turnaround times.

The basis for the STM project was the MONALISA 2.0 project, which defined STM, assessed the strengths and weaknesses of the current maritime ship and transport systems, operations, and interactions, and defined a target concept and key performance indicators for four STM strategic enablers, which are

1. "Voyage management services, which provided support to individual ships in both the planning process and during a voyage, including route planning, route exchange, and route optimization services" [9];

2. Flow management services, which supported both onshore organizations and ships in optimizing overall traffic flow through areas of dense traffic and areas with particular navigational challenges;
3. Port collaborative decision making (Port CDM) services, which increased the efficiency of port calls for all stakeholders through improved information sharing, situational awareness, optimized processes, and collaborative decision-making during port calls;
4. Sea system wide information management (SeaSWIM), which facilitated data sharing using a common information environment and structure (the Maritime Cloud). This ensured interoperability of STM and other services.

In addition, MONALISA 2.0 has developed and created a network of interconnected simulator centers in a number of EU countries. This network enabled testing of STM in complex traffic situations, as well as other functions, like SAR, as a safer alternative to live testing [9].

4.3 Examples of Implementing e-Navigation

As already shown previously, several countries have developed their national strategies based on the IMO's e-Navigation development and the approved SIP. As described in Chapter 1, the SIP is mainly a guide for all stakeholder to raise awareness of specific user needs related to e-Navigation and prioritized solutions to address user needs in a cost-effective manner. This work has to be done in a harmonized and standardized manner.

4.3.1 Canada

The Canadian e-Navigation model may be seen as an example of a national-level e-Navigation structure that could potentially be used by other countries. Canada began by adopting a national vision for e-Navigation and a concept of operations (see Figure 4.1). It then established a governance structure composed of a national committee and regional committees. All committees are cochaired by a Coast Guard director and a marine stakeholder representative. Committee membership is comprised of federal departments responsible for the provision of information to mariners, ports,

Data: Canadian Coast Guard, 2016; NRCAN; EC; Federal Register 1995 Coordinates and NOAA.

Figure 4.1 Context of marine navigation in Canada. (Source: Canadian Coast Guard.)

pilot associations, marine companies, marine industry associations, ferry companies, and other relevant entities. This governance structure allows the committees to take advantage of regional initiatives to implement e-Navigation in a nationally consistent way.

The concept of operations is based on voyage phases (preparation of the sailing plan, voyage, and postvoyage analysis) and aims to provide electronically the correct data during each phase. Identification of the correct data is based on user requirements. Mariners are extensively consulted to provide their input at each phase of a voyage. For example, a user's needs matrix was developed to identify the information that mariners would require to prepare a sailing plan in their respective regions. Mariners also prioritized this information so that official data providers could emphasize their work on the most important data. This matrix led to the creation of the Maritime Information Portal, which facilitates the retrieval of data for each specific region.

This e-Navigation Maritime Information Portal is primarily for use by mariners and the commercial shipping industry, including fishermen and pleasure boaters, who will find the reliable and regularly updated information they need to navigate safely.

Before their trip, mariners will be able to select a specific region and obtain information [8] such as

- Marine weather forecasts and conditions, including buoy and station data;
- Water level forecasts and observations, including tides and current tables;
- Current hazards and warnings, including notices to shipping (NOTSHIP), notices to mariners (NOTMAR), shoals, and the status of aids to navigation;
- Nautical charts and publications as well as chart corrections;
- Status of buoy tending operations;
- Vessel traffic services, including Canadian Coast Guard publications and applicable regulations [10].

A similar approach was used to provide data during the voyage phase. Given that the Canadian Coast Guard is the federal organization responsible for the AIS network in Canada, it was agreed upon by other federal departments to use this network, whenever possible, to provide AIS ASMs. A user's needs matrix containing details of each AIS ASM was developed and distributed to regional e-Navigation (ENAV) committees who provided it to their members. More than 200 mariners and shipping companies completed the matrix and results were used by official data providers as a guiding element for their respective work plans.

Broadcasts of AIS ASMs were initiated in the Seaway and St. Lawrence River in the early mid- 2000s but were not expanded to other regions mainly because international standards for the display of information on ships' equipment were not yet developed. The Seaway and St. Lawrence River pilots were pioneered in this domain using either an ECS or a PPU to display AIS messages on board. In 2009, the IMO approved new regulations for the mandatory carriage requirements of ECDIS (MSC 86, 2009) with a timetable extending until 2018.

This new IMO requirement was linked to e-Navigation information displayed on board and Canada took advantage of this opportunity to reactivate its test beds on AIS ASM. With the collaboration of the Canadian Hydrographic Service (CHS), Environment Canada and Climate Change (ECCC), and the British Columbia Pilotage Authority, the Coast Guard initiated a test bed in the western region on a Met/Hydro AIS ASM. To conduct this test bed, the Coast Guard collected the Met/Hydro data from the databases of CHS and ECCC, converted it to the proper AIS format, and broadcast it from one or several Coast Guard AIS stations. Pilots were consulted to determine the distance from which they wanted to start receiving the data on board. A distance of 30 NM was initially proposed for the trial.

The test bed was conducted in compulsory pilotage areas, and pilots contributed to collect information for display onboard ship equipment. Feedback received from pilots was very positive and the test bed will be expanded to other Canadian regions with the objective to operationalize the AIS ASM communication of information.

Canada also envisions that it will begin testing AIS AtoN in its own waters. Due to the size of the country (the coastline is approximately 243,000 km), the number of aids to navigation deployed (more than 17,000) and the communication systems needed to cover this huge area, Canada initiated an international survey on AIS AtoN in the fall of 2016 to learn from the experiences of other countries. Results showed that the majority of countries were still in a trial phase with this technology; in some cases, benefits were clearly demonstrated for users whereas in other cases, outcomes were mixed.

Based on the information collected from other countries, the Canadian Coast Guard developed criteria and identified situations where the utilization of AIS AtoN may provide an added value. These criteria and situations were presented to mariners for comments and were well received overall. Test beds will be initiated in different areas in the coming months.

4.3.2 The United States

Another example is the United States (U.S.), who has been involved in the development of the e-Navigation concept since its beginnings. Through participation at IMO, IALA, IHO, and various other bodies, the United States has been a supporter and contributor to e-Navigation development. In 2011, in addition to participating in the development of the e-Navigation concept, the main U.S. federal agencies responsible for e-Navigation decided to get ahead of concept development and see what could be done to begin implementing e-Navigation capabilities in advance of the finalization of the IMO e-Navigation SIP. This effort was intended to work in parallel with international e-Navigation concept development, and as appropriate to inform concept development through U.S. participation in IMO and IALA work.

The logical place for this effort to begin was with the U.S. Committee on the Marine Transportation System (CMTS), an existing interagency coordinating body. CMTS is a federal interdepartmental committee chaired by the secretary of transportation and whose purpose is to create a partnership of federal departments and agencies with responsibility for the MTS. The CMTS is directed to "ensure the development and implementation of national MTS policies that are consistent with national needs and to report to the president its views and recommendations for improving the MTS" [11]. The act establishing the CMTS provided that its purpose includes

- Assessing the adequacy of the marine transportation system (including ports, waterways, channels, and their intermodal connections);
- Promoting the integration of the marine transportation system with other modes of transportation and other uses of the marine environment;
- Coordinating, improving the coordination of, and making recommendations with regard to federal policies that impact the marine transportation system [11].

The CMTS organization (see Figure 4.2) consists of cabinet secretaries and leads of independent agencies and White House

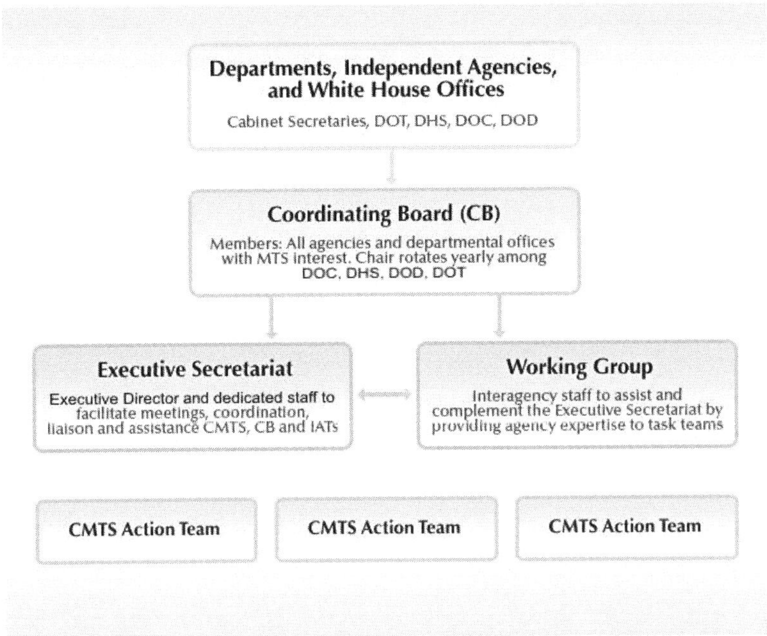

Source: US Committee on the Marine Transportation System (CMTS), http://www.cmts.gov/About/Organization.aspx

Figure 4.2 CMTS organization. (Source: U.S. Committee on the Marine Transportation System (CMTS), http://www.cmts.gov/About/Organization.aspx.)

offices with a role in the marine transportation system. Coordinating board representatives are senior members of the member agencies who set high-level priorities and work plans. The CMTS is supported by an executive secretariat to facilitate the work of the CMTS. A working group consists of a variety of representatives from member agencies who provide agency expertise to assist the executive secretariat. Interagency action teams (IATs) are established to work on specific areas of interest where focused, interagency coordination is needed. Upon development of the U.S. e-Navigation strategic action plan, an e-Navigation IAT was established; further details about it follow.

The lead U.S. agencies in the development of what became the e-Navigation Strategic Action Plan were the U.S. Coast Guard (USCG), U.S. Army Corps of Engineers (USACE), and the National Oceanic and Atmospheric Administration (NOAA). A small team led by representatives from these agencies created a draft that was submitted to the CMTS Coordinating Board and approved in February 2012.

The e-Navigation Strategic Action Plan (e-Nav SAP) established a vision for U.S. e-Navigation implementation:

> To establish a framework that enables the transfer of data between and among ships and shore facilities, and that integrates and transforms that data into decision and action information [12].

The ultimate goal for e-Navigation efforts in the United States is to "use timely and reliable information to make the U.S. Marine Transportation System operate better" [12] and defined "operating better" as safer, more secure, more reliable, more efficient, and more integrated.

The e-Nav SAP focuses on several areas:

- *Integrating systems:* Through a framework of "data standards, information content, delivery means, and decision support applications, ... improve the connectivity of authoritative data sources, eliminate redundant data, and create real-time access" by authorized users [12].
- *Seamless data exchange:* "Achieved through improving the federal government's management of its data, and through the federal coordination of data submission requirements. Navigation information will automatically be pushed to mariners and shore users and will be delivered when it is needed" [12].
- *Human-focused interfaces:* "Configurable interfaces based on user and situational need. This will require system manufacturers, mariners, and offices ashore to collaborate on succinct, efficient displays and methods for displaying dynamic data and information" [12].
- *Decision-focused information:* "Extending system capabilities to include more procedural and decisional functionality as opposed to simply data consumption and display. The government, working with industry and capability providers (e.g., equipment and software manufacturers), will help develop and codify the necessary procedures and decisions to be applied across all transportation systems" [12].

- *Improved connectivity:* "Using federal assets such as sites, bandwidth, and authorities, to see that existing U.S. communications infrastructure fully serves the critical marine transportation zones. High bandwidth, reliable connectivity, and affordability are the goals" [12].
- *Interagency information coordination:* "Increased coordination among agencies, including data interoperability initiatives of the president's ocean policy initiative, will be used to provide consistent data, consistently formatted, and provided via common delivery systems" [12].

To carry out the e-Nav SAP, the CMTS established the e-Navigation Integrated Action Team (e-Nav IAT, renamed the Future of Navigation IAT (FutureNav IAT) in 2015). According to its terms of reference, the objective of the FutureNav IAT is to "facilitate the coordinated and integrated collection, processing and dissemination of navigation data and information to provide services to stakeholders, eliminate duplication, and enhance the reliability, efficiency and safety of our waterways and ports" [12].

FutureNav IAT activities include:

- Identifying existing e-Navigation capabilities including e-Navigation-related test beds and pilot projects;
- Identifying the range of e-Navigation users;
- Enhancing e-Navigation systems interoperability;
- Facilitating the expansion of e-Navigation to achieve coverage throughout the inland, coastal, and offshore regions of the marine transportation system;
- Evaluating the appropriate implementation of e-Navigation systems and the proper mix of e-Navigation systems and traditional aids to navigation and navigation services;
- Clarifying roles and responsibilities of government, nongovernmental organizations (NGOs), technical standards organizations, and industry;
- Aligning U.S. and international e-Navigation efforts.

Upon establishment in 2012, the FutureNav IAT developed its first work plan. The work plan included tasks that were primarily focused on gathering information about current e-Navigation-related activities in the United States and internationally, identifying stakeholders, identifying related legislation and regulations, conducting outreach to identify stakeholder requirements, and identifying priority areas for specific work items.

Specific outcomes of the first work plan included

- Development and approval of a CMTS Coordinating Board Resolution on the use of S-100 for U.S. e-Navigation implementation;
- Compilation of an inventory of existing e-Navigation-related capabilities;
- Compilation of an inventory of e-Navigation-related legislation and regulations;
- An online survey to collect public input on e-Navigation requirements and priorities;
- A series of public meetings to present the e-Navigation concept and gather public input on concerns, requirements, and priorities for future navigation efforts.

In 2015, the FutureNav IAT completed its second work plan building on the efforts completed in the first plan. This work plan includes more specific work items based on the priorities identified through the efforts of the first work plan. Specific task items included

- Development of the enhanced Marine Safety Information (eMSI) concept to harmonize the dissemination of navigation safety information, including input to developing S-100 product specifications;
- An effort to harmonize the identification of waterways to facilitate consistent correlation of information (e.g., MSI, charting features, waterway infrastructure) among the different entities responsible for that information;
- Development and maintenance of an outreach and engagement plan that targets outreach efforts, such as conference par-

ticipation, workshops, and publications with specific task items in support of the e-Nav Strategic Action Plan.

One of the underpinnings of the e-Navigation concept is the ENC, which serves as the base navigation layer that will be augmented with additional layers of information within the e-Navigation environment. To better support e-Navigation and the mandatory carriage requirements for ECDIS for SOLAS class shipping, NOAA initiated an ENC First production process. Currently, NOAA maintains a portfolio of over 1,000 raster navigational charts and over 1,000 ENCs. Prior to moving to a single production system, these products were maintained by two entirely separate teams. ENCs were primarily created and maintained by copying information from compiled raster charts. The ENC First production process shifts the focus to making the ENC the primary product, which means it will be the first to receive chart updates.

Another component of ENC First is an effort to enhance ENC data to better support mariners. This includes exploring the use of metric depth contours, harmonizing ENC scales within the wider hydrographic community, and releasing new large-scale coverage as an ENC-only product.

As with any major undertaking involving various interagency efforts and diverse stakeholders, not to mention ambitious goals, there were many challenges to be faced. Primary among them are the number and diverse responsibilities of the multiple U.S. agencies involved in maritime affairs. This challenge was met through leveraging the existing coordination efforts of the CMTS and through careful review of each agency's responsibilities and limitations.

Another challenge was the lack of comprehensive information on current e-Navigation-related capabilities among the involved agencies and in the navigation industry. This was addressed through the efforts of the first FutureNav IAT work plan, which included completion of a capabilities inventory and stakeholder identification and outreach efforts.

Finally, upon completing the inventory efforts, many gaps in capabilities were identified, particularly relating to e-Navigation infrastructure (e.g., AIS network, communications infrastructure, and data framework). Identification of these gaps and assessment of the priority in addressing them were the basis of the development of specific task items in the second FutureNav IAT work plan.

Since the inception of the e-Navigation concept, the United States has been involved in international efforts to develop and implement e-Navigation. In parallel with this international engagement, the United States has also worked to implement e-Navigation concepts through tangible efforts under an overarching e-Nav SAP. These efforts were aimed at implementing e-Navigation by building on existing capabilities while maintaining alignment with international developments. Many challenges have been faced, and many remain, but the United States is committed to the continued development and implementation of e-Navigation to improve navigation safety, efficiency, reliability, and protection of the environment.

4.3.3 Norway

Norway coordinated the IMO e-Navigation correspondence group from the establishment in 2009 until the delivery of the main output of the work, which was the finalization of the IMO e-Navigation SIP adopted by IMO in 2014. In addition, Norway chaired three IMO working groups on e-Navigation that were related to safety, communications including SAR, and training and human element issues.

Norway has also coordinated further IMO work associated with harmonized display of navigation equipment on board, and chaired an IALA working group on the further development on MSPs. The IALA working group focuses on a guideline for practical and operational implementation of MSPs. Since the delivery of the SIP to IMO, Norway has initiated and participated in several e-Navigation test beds and national and international projects. An example of a Norwegian e-Navigation test bed is automated ship reporting transferred from ships to the Norwegian Single Window system (Figure 4.3). Another is about crucial navigational information provided from maritime authorities to a ship's ECDIS about sea areas that should be avoided or paid special attention due to recent ship accidents, AtoN out of service, oil polluted area, unidentified drifting object, and so forth. A third example is about route exchange information provided between a ship and VTS. More about the test beds are detailed below.

e-Navigation solution 2 consists of two integrated parts. The first part is the automatic collection of data on board and its preparation in the correct format for transmission to the single window application ashore. The second part, which is presented Figure 4.3, is the distribution of a ship's information to the relevant shore parties, such as maritime authorities and customs police defence, via the single window. These two parts are integrated by the communications system.

Another major focus is related to e-Navigation solution 5, on which MSPs are the core of this solution. Projects carried out focusing on the MSP work have been routinely exchanged between VTS and ships, route optimization and electronic request of traffic clearance for a ship accessing a VTS area, which are linked to MSP 3 TOS, and maritime safety information provided from VTS and displayed onboard on a graphical display such as ECDIS. The same information is also displayed on graphical displays at the VTS center. This system supports MSP 5 MSI service. A national annual e-Navigation conference has been established in a close cooperation with users, service providers, research institutes, and manufacturers.

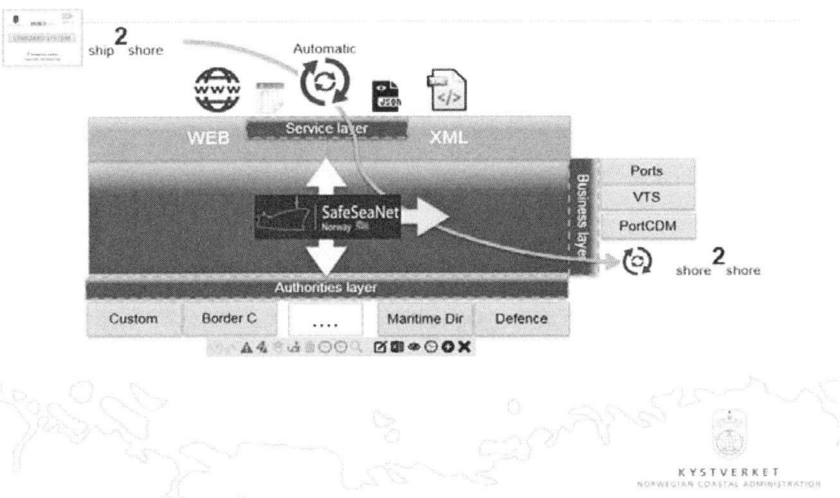

Figure 4.3 The Norwegian Single Window System, supporting authority and business stakeholders with ship-related information. (Illustration: Jarle Hauge.)

In 2016, the Norwegian Parliament approved a white paper presented by the Ministry of Transport through the government titled Preventative Maritime Safety and Preparedness against Acute Pollution [13]. The white paper dealt with preventative maritime safety and preparedness for acute pollution in Norwegian waters and provided a description of traffic forecasts in Norwegian waters and analysis of the risks associated with maritime transport. Then it presented appropriate measures to ensure a high maritime safety level and good preparedness for acute pollution in the years ahead. e-Navigation was one of the measures and the ongoing work on e-Navigation was included in the governmental strategy.

The white paper stated that relevant information, such as navigational warnings, waters regulations, and information about weather, waves, and current conditions, should be easily accessible and presented in a more user-friendly way for navigators. Furthermore, the interaction between navigators, pilots, and VTS centers should be strengthened through new services.

> e-Navigation services will reduce the possibility and consequences of human error, including reducing the complexity of the bridge crew workloads [13].

e-Navigation development and testing of new e-Navigation services has already been underway. The government facilitated cooperation between Norwegian maritime authorities, shipowners, and equipment and service providers with a view to developing new services.

Such cooperation projects fitted well with the government's goal of increasing digitization in public sector and might contribute to the development of technological solutions and services that also will be of interest internationally.

Furthermore, the Norwegian strategy also includes the development of dynamic risk monitoring. VTS contribution to maritime safety can be strengthened further by developing systems for dynamic risk monitoring. Figure 4.4 shows one of the NCA's test bed sites. The purpose of the systems is to identify situations that may develop into accidents as early as possible, which is linked to MSP 2 NAS and MSP 3 TOS.

Dynamic risk monitoring is an automated system that analyzes vessels' sailing pattern in real time based on maritime traffic data

Figure 4.4 The Automatic Information Management and Monitoring Centre and VTS Training Centre at the Norwegian Coastal Administration in Haugesund, Norway. (Photo: Anne Grethe Nilsen.)

available today. This is combined with a static analysis of vessels of risk, based partly on the number of passengers, type of cargo, or earlier recorded events. The system should identify risk vessels, deviations, and potentially dangerous situations as early as possible and enable them to intervene early in order to avert accidents. It may also be appropriate to develop a system that gives automatic notice to the ship by electronic means. Much of the technical infrastructure required is already in place, including systems for traffic monitoring and maritime communications.

Graphical prediction of wave and current forecasts is a project that has been ongoing for several years due to the vast Norwegian coastline. This project supports MSPs 14 and 15, Meteorological Information Service, and Real-Time Hydrographic and Environmental Information Services.

The wave models provide forecasts for wave height and direction up to 48 hours in advance. A summary of the wind strength and direction is also provided.

The wave and current notification service makes it easier and safer for passenger vessels to navigate in exposed coastal areas. This service has been developed in cooperation with users and other Norwegian entities, and it is continuously updated based on the needs of the users. In addition, information from local wind sensors along the coastline is delivered to a broad number of stakeholders via an application.

A new project is planned to equip aids to navigation with sensors for monitoring the system and to communicate with ships in case of abnormalities. This project supports MSP 5 Maritime Safety Information Service and MSP 10 Nautical Publications Service.

The information portal BarentsWatch [14] was launched in 2012 and is being developed incrementally. The portal has information about topics such as the climate and environment, marine resources, oil and gas, and maritime transport and maritime law, among other things. There are also map services, an overview of ports, and news from about 25 national partners. Several new e-Navigation services are planned in this respect.

The Hydrographic Service of the Norwegian Mapping Authority is responsible for preparing and updating nautical charts and publications, such as fairway descriptions for all marine and coastal waters in Norway, including Spitsbergen as well as other polar waters. They are about to transfer all analog products to digital services. This work supports the further development of MSPs.

In Europe, the e-maritime concept aims at promoting the competitiveness of the European maritime transport sector and a more efficient use of resources through better use of ICT tools. Modern ICT systems provide undeniable benefits that are not available through paper-based information such as automated information verification and analysis, processing of data and optimization routines, and easy sharing of information already submitted or stored. Solutions to allow reuse of data already submitted, remove unnecessary reporting obligations, and optimize port and ship processes will be discussed in the context of the Digital Transport and Logistic Forum. Norway is one of the members of this forum.

"The Forum, set up by the [European] Commission, will gather experts, business operators, and policy makers in order to identify the needs and to prepare for common actions at the EU level for improving freight transport and logistics with more efficient use and reuse of digitalized information currently produced and stored by many different stakeholders" [15]. Details about the EUs implementation of the Maritime Single Window and the European ship reporting system SafeSeaNet was presented in Chapter 2.

4.3.4 Australia

In November 2006, the AMSA hosted an e-Navigation workshop and invited key Australian maritime stakeholder groups to participate. The workshop agreed that e-Navigation should ensure that mariners' needs are addressed. Vitally, it agreed that "Human factors are a major focus in the development of e-Navigation, in terms of solving existing human factors problems, improving navigation situational awareness, and also to help avoid unintentional new human factors" [16].

Since then, Australia has focused on improving the usability of navigational systems. A second workshop was held in 2013 to further develop this concept (earlier that year, the NI held a workshop at the World Maritime University in Malm, Sweden, to kick-start the process). IMO member states, international human factor experts, academics, experts from other industries, and equipment manufacturers participated in the Australian workshop.

The workshop laid the foundation for what was to become a first-ever IMO guideline on the design of navigational systems. Australia, along with the ROK and Japan, led the development of an IMO guideline titled *Software Quality Assurance and Human-Centred Design for e-Navigation* [17]. The guideline is a nonregulatory approach to support the principles of SOLAS V/15 (which deals with the principles of bridge design), and is primarily intended for developers and testers of e-Navigation systems.

The basic premise of HCD is that systems ought to be designed to suit intended users and their tasks rather than requiring users to adapt to a system. The software quality assurance aspect offers guidance on defining and testing software quality to ensure high quality and stable software.

In 2010–2011, IMO chose IHO's newly established S-100 Geospatial Information Registry as the agreed CMDS for e-Navigation. The S-100 ecosystem is eminently suitable for developing data models (or products) that serve as templates for the exchange of e-Navigation-related information. Through a combination of formal submissions and personal networks, Australia was instrumental in influencing the IMO's choice of S-100.

AMSA's Under Keel Clearance Management (UKCM) system [18] has provided an e-Navigation service since 2011 for large ships transiting Torres Strait in Australia's north. It is a current-generation system for enhancing the safety of ships whose keel may be close to the seabed when transiting the region. Such ships face many additional challenges in navigating Torres Strait, due to the numerous reefs, complex tides, and strong tidal streams that characterize the region.

The UKCM system uses known ship stability information together with hydrodynamic modeling to predict the dynamic motion of the ship (squat and heel). When coupled with the latest bathymetry and observed environmental data from tide, tidal stream, and wave sensors, the system is able to compute an accurate estimate of a ship's UKC, in real time.

The system allows coastal pilots and ship operators to plan and monitor safe and efficient passages of deep-draught ships transiting Torres Strait. The "collection, integration, exchange, presentation and analysis" of UKC information [19] fits well with the IMO definition of e-Navigation. The efficiency benefits that will flow from the knowledge of accurate UKC (by potentially permitting deeper-draught ships to transit) are being studied at the time of writing.

Traditionally, UKCM systems have been used in ports to maximize cargo uplift and throughput, thereby ensuring safer resultant transit of bulk carriers, container ships, and tankers. This is certainly the case at many of the major ports in Australia. AMSA's UKCM system in Torres Strait has extended this notion to include a coastal waterway used by transiting vessels.

In 2015, the IHO agreed that the AMSA and the Australian Hydrographic Service (AHS) would lead international efforts to develop an S-100-based product specification for displaying UKCM information on ships' navigation systems. Currently, it is difficult for ships' officers to fulfill their passage planning and berth-to-berth route monitoring SOLAS obligations due to a lack of information about UKCM systems in the areas in which they operate (primarily ports, but also some waterways such as Torres Strait and potentially SOMS). Work on this project has commenced and will gather momentum in 2017.

Since its advent in the late 1990s, AIS has been a stand-out success. Originally meant for use in a ship-to-ship mode for collision avoidance and as a means for littoral states to obtain information about a ship and its cargo (i.e., as a VTS tool), its use has grown immensely. Authorities now use AIS to support their aims and charters. Uses include:

- Search and rescue;
- Ship operations;
- Salvage and intervention;
- Investigations and prosecutions;
- Port state control;
- Compliance with pilotage requirements;
- Environment protection;
- VTSs;
- Planning of navigational services (e.g., ships' routing measures);
- Monitoring of (and use as) aids to navigation;
- Strategic planning.

With regard to the ITU, Australia has played a leading role in seeking approval to establish the VDES. The VDES concept addresses the need for additional capacity for digital data exchange in an e-Navigation world. In 2015, Australia, along with commercial interests, conducted sea trials to prove the VDES concept (for more information about VDES, see www.amsa.gov.au).

Australia is modernizing its AIS shore station network to use software-defined radio. Australia's AIS service manager is also VDES-ready and complies with the VDES standard as it exists today. Australia continues to prepare for and support the introduction of VDES through the introduction of new hardware and software in its shore station network.

Regarding PNT, Australia advocates that, as far as is practicable, there is a consistent approach to the provision of PNT for marine navigation worldwide. From the outset, Australia has cautioned against the use of differing systems for PNT in different regions. This would potentially create new risks for navigation, as mariners will need to change their methods and practices when traveling between regions. The IMO agreed to several e-Navigation outputs in 2015 that will progress between 2016 and 2019, and according to AMSA's website, AMSA will continue to contribute to these IMO endeavors focusing on the human element and the provision of appropriate PNT.

Australia supports "the inclusion of information available on board a vessel that can be sensed, particularly when navigating in coastal areas. Sensed information onboard includes visual bearings, radar ranges and bearings to radar conspicuous features, water depth, radio direction finding and ranging signals, and potentially even automated sextant angles to celestial bodies. Australia is of the view that as e-Navigation evolves, the effective use of vessel-sensed-information should be accommodated to enhance system resilience, robustness, and mariner confidence in navigating their vessels" [20]. In parallel, the AMSA has been actively engaged at various national strategy and working groups with the aim of promoting a common national, multimodel approach to the provision of resilient PNT.

Australia continues to work on improving the design of navigational equipment in terms of developing an S-Mode of operation. Australia is informally coordinating the work of a coalition of international experts and aims to develop a draft IMO guideline on the S-Mode of operation for navigation equipment such as ECDIS by 2018. "S-Mode involves the standardisation of design for displays, interfaces, and functionalities. It aims to improve the ease of use of [ECDIS] and other navigation systems" [20]. The draft IMO guideline will be based on ISO standards and user needs developed for e-Navigation. The need for S-Mode was first raised by the NI and the International Federation of Ship Masters Association (IFSMA) in 2008.

In a move to reduce manual ship reporting and maximize the use of AIS technology, Australia introduced changes to the reporting mechanism of the mandatory Australian Ship Reporting System (AUSREP) in 2012. AUSREP, which involved the lodgment of sailing plans and regular position reports by Inmarsat, HF, satellite telephony, or other means, was amended to the Modernized Australian Ship Tracking and Reporting System or MASTREP. The centerpiece of MASTREP is the use of AIS data to satisfy AUSREP reporting requirements.

This more streamlined, automated approach has alleviated the obligation for ships to lodge daily position reports. In addition, MASTREP data is more accurate and offers the following benefits:

- Increase in the timeliness and accuracy of data collected from a much larger number of vessels in the Australian Search and Rescue Region (SRR);
- Provides AMSA with the capability to significantly improve its SAR, pollution, and marine casualty responses through more effective management of incidents and resources.

A goal of e-Navigation is to provide better designed harmonized information in electronic formats in a seamless and efficient manner to navigational systems on board (see Figure 4.5).

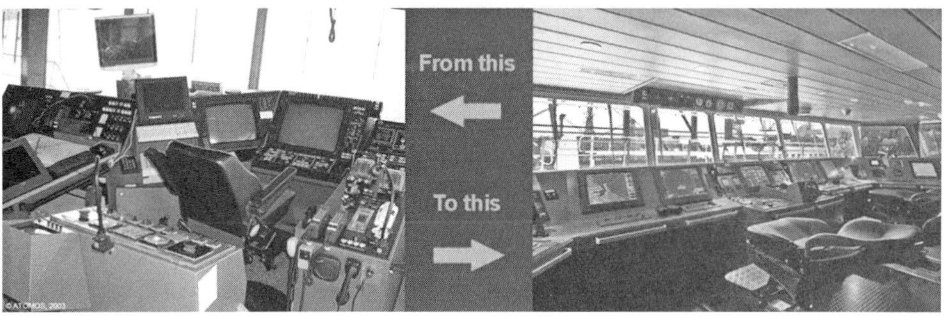

source : www.amsa.gov.au

Figure 4.5 A before and after e-Navigation example according to AMSA, (Source: www.amsa.gov.au.)

Like other areas, e-Navigation is the maritime world's response to the increasingly rapid digitalization of human activity. Ashore, e-Navigation aims to streamline the way maritime authorities, agencies, and other stakeholders gather and exchange information.

4.4 Expectations of Maritime Equipment Manufacturers

Since the inception, industry has actively followed the development of the IMO's e-Navigation initiative. Throughout the last few years it became obvious that the reaction of the maritime industry is not necessarily aligned. This is not surprising, as the maritime industry is not necessarily homogeneous. In order to describe an industry view on e-Navigation, it is important to first look at the different sections of the industry to be aware of the different interests and viewpoints. This will help to identify and describe the common views, as well as the differing aspects.

The maritime industry is widespread, which requires an examination of many different components in order to clearly describe this industry. However, the focus below looks at two key sectors that have a high impact on the industry view of e-Navigation: the shipping industry and the ship supply industry.

One aspect within this framework that needs to be looked at separately in terms of the relationship to e-Navigation is the shore-side sections of the maritime industry, with key players such as shore-side authorities and their suppliers.

The first part of what we consider the maritime industry can be categorized as the shipping industry. A definition of what the shipping industry is can be derived from the IMO [21]:

> Its main role is to create a regulatory framework for the shipping industry that is fair and effective, universally adopted and universally implemented.

In other words, its role is to create a level playing-field so that ship operators cannot address their financial issues by simply cutting corners and compromising on safety, security, and environmental performance [21].

In this view, the shipping industry is focused on ship operators. The players in the field have formed associations, such as

InterTanko, Cruise Lines International Association (CLIA), InterCargo, and InterManager.

A second important sector of the shipping industry are the shipowners, who are not mentioned in the IMO definition. But when looking at the self-declaration of ICS, this aspect of the shipping industry becomes obvious [22]:

> The International Chamber of Shipping (ICS) is the principal international trade association for the shipping industry, representing shipowners and operators in all sectors and trades
>
> Both groups, ship owners and operators, are in essence responsible to ensure safe and efficient transport of goods internationally by seaways, in a most cost effective way for them to stay competitive.

The enormous economic pressure on the shipping industry, as defined above, especially after the economic crisis at the beginning of this century, manifested itself in a very conservative approach toward new development. While the shipping industry is always considering how innovation can help to increase efficiency and with that reduce costs, new developments like e-Navigation are looked at carefully, perhaps with the concern that other players in the maritime industry are attempting to introduce additional expenditure without necessarily providing a return on investment. Here lie the reasons for the hesitation on the ECDIS mandate, but also the careful consideration, and at times criticism, of the e-Navigation initiative.

What has transpired in the e-Navigation discussion is that the shipping industry, while generally appreciative of the e-Navigation initiative, is looking for a very clear cost-benefit analysis of any parts of this initiative and will only support further development if this analysis shows positive results for improved efficiency. Even more so, if coming from IMO, it needs to ensure that the initiative is compliant with IMOs role "to create a level playing-field", as the President of CIRM, Michael Bergmann, describes it. Therefore, any implementations need to be adapted for all players to ensure they are not supporting only a few while penalizing others. In other words, the shipping industry is looking for implementations that are enforced by all—not just some—flag states.

The second major sector of the maritime industry is the ship supply industry.

In this sector, one finds shipyards, electronic equipment, and software manufacturers, as well as service organizations for ship supplies and repairs. Important associations representing this sector of the industry include CIRM, Community of European Shipyards Associations (CESA), and International Shipsuppliers & Services Association (ISSA).

The general interest of this sector is to supply the shipping industry with the equipment and services it needs to do its job. In contrast with the more conservative approach of the shipping industry described above, this sector is always seeking innovative ways to support its customer base, creating new products and services that are either mandated by IMO to ensure market penetration or provide a return on investment attractive to its customers. Ideally, both conditions should come into play.

Therefore, the ship supply industry often looks at the e-Navigation initiative as a good vehicle to introduce new innovative products and services providing mutual benefit for both maritime industry sectors. It is supporting—sometimes with high costs—e-Navigation test beds to validate ideas that can be realized and put into production, and to prove that the solutions provide the necessary benefits for the shipping industry, so the net result of the cost-benefit analysis is positive.

It is important to this sector is that developments in e-Navigation can be turned into products and services that may be brought to market in the foreseeable future. It is not interested in long-reaching theoretical exercises that do not lead to implementable results. It is pushing for the implementation of e-Navigation solutions rather than even more test beds and further research and analysis. It advocates phased approaches, which manifest in new products and services but do not close the door for rapid innovations in future.

It is expected that if e-Navigation does not transition from theory to practice in the near future, the ship supply sector may lose interest. But, at the same time, it is already visible that the ship supply industry is actively implementing solutions that are in effect attributed to e-Navigation. Implementations such as UKC solutions, large integrated displays, innovative route planning solutions, and ship reporting systems are just examples of this new reality created by this sector of the maritime industry. It will be the task of IMO and other players within the e-Navigation initiative, such as IALA or IHO, to ensure those solutions become integrated and linked into the overall concept of e-Navigation.

The shipping industry is aware that it needs to find innovative solutions and as such is supportive of the IMO e-Navigation initiative. But any change this might create in the finely balanced shipping industry environment is looked at very carefully, and at first may been seen as a potential threat. One could see this as a conservative approach towards e-Navigation, but it should not to be seen as opposition to the initiative.

The ship supply industry tends to support e-Navigation as an opportunity to create new and innovative products and services for the shipping industry, which must pass the critical eyes of its customers who are looking for a positive result of any cost-benefit analysis. This sector is already developing and offering solutions built on the vision developed by the e-Navigation initiative.

Both sectors of the maritime industry are looking toward regulators and standardization bodies to create the necessary framework to enable the solutions to be globally adopted. The maritime industry is international by nature. As such, any solutions need to be globally acceptable, keeping in mind that while regional conditions may need region-specific solutions, the maritime industry is too small to allow for those that are isolated and noncompatible.

The IMO e-Navigation initiative is not only ship-focused, while most of the other IMO activities are mainly related to ships. The organization realized that improvements in safety and efficiency of maritime trade needed to include the shore side. As such, the three focus areas identified by IMO are the ship, the shore, and the communication between the two, and between players within one of these areas. In general, on the shore side, two types of stakeholders are involved: governmental administrations and shore-side maritime industry players.

Any costal state has obligations to protect, maintain, and manage territorial waters. Key elements of these obligations are to

- Protect borders and ensure any activities within territorial waters are within the laws;

- Set necessary local regulations—in accordance with international regulations as set by the IMO and others—to ensure safety and environmental protection within respective waters;
- Ensure that different usage of territorial waters by different players are compliant to each other and appropriately coordinated, for example by adequate spatial planning of the sea area in question;
- Support economic growth by managing areas of congestion.

While these are all sovereign responsibilities of public administrations, regional coordination and collaboration have been established throughout the world. Examples of this cooperation are EU activities or multilateral coordination as in the Straits of Malacca and Singapore (SOMS). In the IMO, different countries are often represented by members of their coastal administrations and as such e-Navigation is in essence initiated and supported by those organizations. Coastal administrations and regional cooperation bodies work in often very different environments, where conditions and constrains may be very different between different regions. Given the local focus and obligation of the involved administrations, it at times is difficult to come to a necessary harmonization of implementations and regulations to allow international maritime trade to work in a somewhat synchronized environment.

One of the key organizations that aims to work collaboratively on this issue is the IALA. Within the IALA's e-Navigation Committee, coastal administrations, as well as industry members, are working together to shape e-Navigation. The main focus of course are the shore-side aspects, but as all the aspects of e-Navigation need to fit and work together, IALA was instrumental in also developing general, holistic aspects of the IMO initiative, like the e-Navigation architecture and its contributions to the SIP.

The challenge IMO has in moving e-Navigation from the conceptual phase to reality is to help local and regional administrations agree on common infrastructure in order to implement MSPs. MSPs are groups of shore-side services that enable ship-side e-Navigation to work. Around the world regional implementations of e-Navigation are in the works, and some, such as UKC systems, are already implemented. Regional test beds are also further validating the concepts. The IMO's further e-Navigation challenge will lie in establishing rules, guidelines, and regulations to ensure compatibility of regional implementations of the general concepts.

The industrial sector of the shore-side stakeholders can be divided in two main categories.

In this sector fall ports with associated services and installations, pilot organizations as well, as VTSs, to name three prominent aspects. The responsible organizations in this section are by nature not globally oriented, but locally focused. However, there is also a lot of collaboration between them and partnerships are not uncommon. Collaboration with shore administrations is also an essential part of oiling the machine of maritime trade in regional and local areas. Important services of e-Navigation, such as automated ship reporting or ship traffic management are finding their ways into national or regional regulations based on international regulations or recommendations. National Single Window or regional Single Window reporting are initiatives within e-Navigation to reduce the workload of the navigator and in turn increase safety. However, e-Navigation services of this stakeholder group also improve efficiency and make better use of assets such as sea areas or port installations. As such, e-Navigation—once implemented—can develop as an essential tool for these players to execute their mission. It is a similar situation for shore administrations, where the difficulty will be to ensure compatibility of locally driven implementations, taking local conditions and constrains into account in terms of implementations in other regions of the world with different needs and desires.

In this category we find equipment manufacturers of tools used in VTSs and other tools necessary to execute the above-mentioned services. Only in rare cases, often with a lot of private initiatives, regional tools are developed and locally adapted without specific impetus to harmonize globally. A good example of this are PPUs, which are tools often created locally by pilots for colleagues in the same association or region. They do not necessarily require harmonization, as they are developed and used locally.

In general, it needs to be realized that besides the above examples, the industry is too small to justify the development of tools, such as VTSs, which are only usable in one region or country.

As well, ship systems need to globally utilize those services, and therefore the installations need to be in a position to receive or deliver data in a harmonized way globally as ships trade around the globe. Given these constraints, the players in this category are eager to be involved in the development of e-Navigation and invest money and energy in test beds to ensure their systems and services can be used as broadly as possible. They organize their involvement to a large degree in what are known as NGOs. These are trade or industry organizations such as the International Maritime Pilots' Association (IMPA), the International Harbour Masters' Association (IHMA), and CIRM. The support of these NGOs are as equally important for the success of e-Navigation as the willingness of coastal administrations to establish the necessary infrastructure.

The shore-side aspects of e-Navigation are driven by local needs, conditions, and constraints even more so than those of the ship side. The obligations of coastal administrations meet the economic and ecological needs of the regional or local industry players and service providers. Harmonizing these locally focused aspects with the global needs of the maritime industry at large is a key aspect—if not the most important one—of the e-Navigation initiative.

The IMO has successfully launched these efforts and created such important artifacts as the e-Navigation SIP, various guidelines, and an overarching architecture, as well as agreeing on a CMDS built on the IHO GI Registry, often referred to as the IHO S-100 Registry.

It is paramount for the success of e-Navigation that the IMO, as well as the NGOs and intergovernmental organizations (IGOs), continue their efforts to move this concept into reality. The maritime industry is international and no local economy can live without it. In the growing connectivity of the world, the harmonization of maritime sea transport is a necessity for the world economy to stay vital and manage the challenges ahead.

4.5 Communications

An essential part of the IMO e-Navigation initiative is communication ship to shore, ship to ship, and shore to shore. Often referred to as the third dimension, e-Navigation will not work without this part. Three aspects are central in this context:

1. The communications industry;
2. The communications equipment industry;
3. The involvement of administrations;

Communication between shore-side installations is well established and doesn't need further review with regard to e-Navigation as far as it relates to the communications industry.

To enable e-Navigation, there is a need for establishing the necessary communication both between ships as well as with shore. The communications industry is increasingly offering solutions for the high seas. SatCom services, not only the classic Inmarsat for GMDSS, but also other communication services to connect ships via satellite with the shore, are becoming available. While prices are becoming more affordable, bandwidth is also increasing. Creative subscription models are being offered to allow customization for the different needs of a shipping customer. But it still has to be noted that such communications can be expensive, so any e-Navigation solutions need to take these costs into account and consider how to reduce bandwidth occupation.

The situation near shore is a bit different. Here the communications industry already offers different solutions. Combinations of different methods, such as third/fourth generation (3G/4G), Worldwide Interoperability for Microwave Access (WiMAX), or even Wi-Fi, are regionally available and are supporting transmissions of higher data volume once the ship is within coverage.

Other communication methods for data exchange, such as AIS-ASM, have been tested in various e-Navigation test beds. The development of VDES is another aspect of communication that has increasing importance in the implementation of e-Navigation. In general, the growing utilization of bandwidth needs to be carefully considered when looking at communication services for e-Navigation. Data transfer needs should not create conflicts with GMDSS or classic AIS needs. The discussion on VDES and the proposed frequency bands aim to reduce interference and ensure that essential services can coexist. This will be a major challenge in the future for organizations such the ITU as well as for the entire communications industry.

Equipment manufacturers build their system to utilize the communication methods available. They cooperate with the communications industry and utilize the directives by organizations such as ITU to provide the necessary tools for smooth and reliable communication. Multireceivers allow seamless selection of the best communication methods based on given characteristics such as cost, bandwidth, and reliability.

Another important aspect the communications equipment industry has to work on is cybersecurity, as described in earlier chapters and detailed in Chapter 1. This is a holistic challenge for both shore-side and ship-side equipment. Authentication and other aspects need to be handled in close coordination with the communications industry as well as with the other equipment manufactures utilizing the connectivity tools.

In order to work correctly, e-Navigation will need shore-side infrastructure and maritime services, which will be either completely provided by shore administrations or at least monitored and regulated by them. This includes the availability of communication means. Infrastructure for VDES, WIMAX, or in some areas even support for wide-reaching G3/G4 coverage will require administrative support and engagement. Here again the involvement of administrations such as the IALA and ITU is essential to ensure communication availability as well as privacy protection as required by local legislation.

Communication is key to e-Navigation. The close cooperation of all involved parties is essential. This aspect often doesn't receive the attention it deserves and needs. While in test beds, communication can be treated with less focus as only a limited amount of data and bandwidth occupation will be needed, but if systems are to be implemented in areas of high traffic—areas where e-Navigation solutions are more necessary then elsewhere—communication can quickly become the bottleneck and jeopardize success. Results of test beds in various regions of the world should be utilized to create prognoses of bandwidth and frequency needs during implementation phases of MSPs.

4.6 The Link

e-Navigation, as often stated, is an IMO initiative. As such, any aspects in the maritime industry correlated to e-Navigation needs to be linked one way or the other with this IMO initiative and the related documentation. But it also has to be realized that e-Navigation is not only tackling what is called front of the bridge, it is also looking into all aspects to improve the safety and efficiency of navigation. This includes aspects that are clearly not done during execution of a voyage or by the OOW.

Furthermore, e-Navigation is by intention highly innovative, and therefore new developments may be linked to the IMO initiative, which has not yet found its way into respective IMO conventions or resolutions. In fact, the IMO is currently focusing its documentation mostly on e-Navigation toward guidelines. These documents are not performance standards, which are endorsed by IMO and strictly need to be followed, but rather they set the scene for developments that don't need to be restricted by the long cycles used by the IMO to process mandatory requirements. This intended flexibility is an essential aspect of e-Navigation. Restricting development within confined areas of type approval and strict and explicit IMO endorsement will eliminate the benefits that e-Navigation could provide.

But providing the necessary freedom for the industry to develop innovative and forward-looking solutions compliant with the IMO e-Navigation paradigm also poses the risk that some may utilize equipment or services in ways that are not within that scope.

The advantages and disadvantages of using performance standards, which take time to revise, and type approval to ensure correct operation of a system, versus the freedom of using guidelines, which have do not stifle innovation and can keep pace with technology changes, are discussed further in the next chapters.

References

[1] IMO, *Revised Guidelines and Criteria for Ship Reporting Systems* (Resolution Msc.43(64)), Document NCSR 3/10, December 21, 2015.

[2] ACCSEAS Final Report, May 13, 2015, available at http://www.accseas.eu/content/download/8190/74102/ACCSEAS%2520Final%2520Report%2520v1.pdf, accessed May 10, 2017.

[3] Motz, F. "German e-Navigation Testbed TICON: Putting IMO Solution S-4 Integration and Presentation of Available Information in Graphical Displays Received via Communication Equipment' to the Test," presented at *e-Navigation Underway North America 2014*, April 4, 2014, available at https://www.csum.edu/c/document_library/get_file?uuid=178efc7f-6292-41b2-a2ca-33899ea6a641&groupId=61938, accessed May 10, 2017.

[4] IMO, *Work Programme, e-Navigation–New Output on Harmonized Maritime Service Portfolios*, Document MSC 96/23/7, February 9, 2016, available at http://www.iho.int/mtg_docs/com_wg/NIPWG/NIPWG2/NIPWG2-33.1_MSC%2096-23-7%20-%20e-Navigation%20-%20New%20output%20on%20harmonized%20Maritime%20Service%20Portfolios%20%28Australia,%20Canada,%20Finlan...%29.pdf, accessed May 10, 2017.

[5] Sesame Straits, *A Brief Overview of the Sesame Straits*, available at http://www.straits-stms.com/Project.html, accessed May 10, 2017.

[6] "Tianjin Port to Establish 'E Navigation' Project," *China Daily*, November 19, 2014, available at http://www.chinadaily.com.cn/m/tianjin2012/2014-11/19/content_18943723.htm, accessed May 10, 2017.

[7] *Buoys Marine Meteorological Observation System—e-Navigation*, available: www.e-nic.cn/product/product1_eng.php?id=1 [authorized access only].

[8] Tunheim, N., "Designing for Healthcare at Sea," in AHO Works Studies 2014–2015, December 2015, available at http://flipbook.cultrixmedia.com/rk-grafisk/aho/aho/aho-works-studies-2014-2015-ver2/#page/1, accessed May 10, 2017.

[9] *Sea Traffic Management Validation Project*, available at http://stmvalidation.eu/#!activity-item/activity-2-voyage-management, accessed May 10, 2017.

[10] Government of Canada, *e-Navigation Maritime Information Portal*, available at http://www.marinfo.gc.ca/e-nav/about-apropos-eng.php, accessed May 10, 2017.

[11] Committee on the Marine Transportation System, homepage, available at http://www.cmts.gov/About/Index.aspx, accessed May 10, 2017.

[12] Committee on the Marine Transportation System, *e-Navigation Strategic Action Plan*, February 2012, available at http://www.cmts.gov/downloads/CMTS_e-Navigation_Strategic_Action_Plan_Feb_2012.pdf, accessed May 10, 2017.

[13] Norwegian White Paper 35, *On the Right Track*, 2015–2016.

[14] BarentsWatch, available at https://www.barentswatch.no/en/, accessed May 10, 2017.

[15] European Commission, *National Single Windows and e-Maritime*, available at http://ec.europa.eu/transport/modes/maritime/e-maritime_en.htm, accessed May 10, 2017.

[16] Australian Maritime Safety Authority (AMSA), e-Navigation workshop, November 2006.

[17] IMO, "Guideline on Software Quality Assurance and Human-Centred Design for e-Navigation," Document MSC.1/Circ.1512, July 13, 2015, available at

http://www.ics.org.ir/Downloads/CLD/News/MSC.1-Circ.1512%20-%20 Guideline%20On%20Software%20Quality%20Assurance%20And%20Human-Centred%20Design%20For%20e-Navigation%20(Secretariat).pdf, accessed May 10, 2017.

[18] Australian Government, AMSA, *Under Keel Clearance Management System Fact Sheet*, available at https://www.amsa.gov.au/forms-and-publications/Fact-Sheets/UKCM_FS.pdf, accessed May 10, 2017.

[19] IMO, *Strategy for the Development and Implementation of e-Navigation*, Document MSC 85/26/Add.1 Annex 20.

[20] Australia Government, AMSA, *AMSA Abroad*, August 2016, available at http://www.amsa.gov.au/forms-and-publications/about-amsa/publications/AMSA-Aboard/2016-Aug/index.asp, accessed May 10, 2017.

[21] IMO, *Introduction to IMO*, available at http://www.imo.org/en/About/Pages/Default.aspx, accessed October 4, 2016.

[22] ICS, *About ICS*, available at http://www.ics-shipping.org/about-ics/about-ics, accessed October 4, 2016..

5

Standards

5.1 The IMO Process

Standards are everywhere in our lives. Without standards, nuts would not fit bolts and it would not be possible to be sure that what we are buying is fit for the purpose.

The World Trade Organization (WTO) encourages standards to ensure that there are no technical barriers to trade, thereby encouraging free markets.

Ever since the Titanic disaster, nations involved with world shipping have strived to ensure that ships are built to agreed-on standards and have some harmonization in their operation and control. What started as an international initiative between shipping nations has become a United Nations specialized agency, the IMO, which currently includes 171 member states. One of the key publications is *SOLAS,* which defines the communications equipment (Chapter IV) and navigation equipment (Chapter V) necessary for each type and tonnage of ship. As an international convention, each country with ships flying their flag (flag state) has to have their own legislation mirroring the IMO requirements and subsequently be responsible to ensure that equipment fitted meets the IMO performance standards referred to in Chapters IV and V. These standards ensure that the equipment defined therein meets minimum performance and safety requirements laid down by the appropriate IMO committee. This can be as simple as the accuracy of a compass or a complex as the ability of a radar to see a small target in bad conditions. They also ensure that the equipment will work in the environment found on ships and will not interfere with each other.

The IMO's range of standards are known as performance standards, defining the minimum performance requirements of navigation aids and communications equipment found on the bridge of ships.

With the introduction of e-Navigation, it is necessary to review existing IMO standards and understand which will need modification so that the specified equipment can function in an integrated and harmonized way.

When a new piece of navigation or communications equipment is identified, often due to new technology, a proposal has to be made to the appropriate IMO committee with a full justification to show the need for a performance standard. If this proposal is approved, then the relevant subcommittee will take on the work of drafting a standard. The working group of the subcommittee will normally have a strong contingent of industry experts who can ensure that the requirements are feasible.

After usually two sessions of the relevant subcommittee, the standard is submitted to the committee for approval and is given a resolution number. The whole process from concept to final approved and published IMO performance standard normally takes five years.

If the new equipment is identified for mandatory carriage on ships, then the equipment will be added to the carriage requirement list in *SOLAS,* Chapter IV (communications) or Chapter V (navigation). A footnote is added referring to which performance standard is appropriate. Footnotes are used because they do not form part of the agreed and ratified text of the convention, which cannot be changed easily, and if the performance standard is changed for some reason, the footnote can be updated without amending *SOLAS.*

Therefore, for every piece of navigation and communications equipment used on a ship's bridge, there is an IMO performance standard.

One of the most important performance standards is A.694(17) *General Requirements for Shipboard Radio Equipment Forming Part of the Global Maritime Distress and Safety System (GMDSS) and for Electronic Navigational Aids*. Note that the number defines the type of resolution, with "A" being an assembly resolution, and the number in brackets is the number of the assembly that the resolution was agreed. This standard includes basic ergonomics, use of controls and indicators, illumination for using the equipment at night, power supply arrangements, environment including temperature and humidity, electromagnetic compatibility and interference, equipment noise (for example cooling fans), safety aspects, service and maintenance requirements, and marking and compass safe distance. This last parameter is important as equipment made of a magnetic material (steel) or containing magnets (some loudspeakers) should not affect the ship's standard magnetic compass, which might be used as a last resort if the power fails.

The advantage of this standard is that it can be referred to in all other bridge equipment standards without having to repeat these general requirements. This means that to add general requirements for e-Navigation, such as built-in integrity testing or interfacing, this is a good place to do it. As a typical IMO performance standard, however, it indicates that a specific requirement should be met without defining test methods. The test methods are left to other organizations who understand testing of electronic equipment to prepare. This is discussed further in Section 5.2 on testing standards.

Standards need to be written with care so as not to stifle innovation. It is important that the requirement of the standard is defined regardless of current technology so that innovators may implement better technology when it becomes available. This is the case, for example, for radar and ECDISs, where careful standards drafting ensured that the change in display from cathode ray tubes (CRTs) to flat panels could be made without rewriting the requirement.

Both Chapters IV and V of *SOLAS* contain a statement saying that equipment fitted to ships to meet mandatory requirements should be type approved by the administration. It would then be the responsibility of the flag state to ensure that the equipment, when fitted to one of their ships, is of a type that has been shown to meet the requirements, typically by testing a sample to ensure the requirements are met, which is called type approval.

IMO standards do not usually contain test methods and so another testing standard is needed. Many administrations used to have their own national test standards, which meant that other administrations did not always accept the approval certificates. Due to the need to reduce barriers to trade from the WTO, and under pressure from the manufacturers, whose type approval costs were increasingly more expensive in order to reach new markets, most testing standards are now internationally agreed on and are published by international standard organizations such as the IEC for electrical/electronic equipment and the International Organization for Standardization (ISO) for mechanical devices. In a few cases, some countries still require their own testing standards to be used but many agreements now exist to recognize agreed-on international standards.

The advantage to industry is that the use of internationally agreed-on standards means that type approval in one country is now often accepted in another without requiring more testing and the accompanying mass of paperwork.

Most developed countries have standards bodies that prepare national standards for all types of products and services: the American National Standards Institute (ANSI) in the United States, the British Standards Institute (BSI) in the United Kingdom, and Deutsches Institut für Normung (DIN) in Germany. International collaboration between these institutes created two central international standardization bodies, one dealing with mechanical standards (ISO) and one dealing with electrical and electronic standards (IEC). Both have central offices in Geneva where the standards are published. These organizations set up committees to deal with specific issues; the relevant committees for bridge equipment on ships are ISO TC8 and IEC TC80.

Each committee is the responsibility of a national standards institute for administrative purposes, such as ANSI in the United States and BSI in the United Kingdom. Membership of a committee is made up from those national institutions who wish to take part and support the work of a committee, and will ultimately vote to support, or vote otherwise, for a standard.

Individual standards are drafted by a working group appointed by the committee, which may also include technical experts or observers as the working group sees fit to consult. Once a project is approved, there are strict time schedules monitored at the central office to make sure that the project runs to time, and the standard is published.

The committees consist of P (participating countries) members from national standards authorities as well as O (observer countries) members from other countries. The P members are often supported by manufacturers accredited by their national authorities as experts on a subject. For example, in IEC Technical Committee 80 (Maritime Navigation and Radiocommunication Equipment and Systems) there are 16 P members and 18 O members.

Agreement at each stage in the process is determined by majority voting by the P members. At some stages, commenting is encouraged with the vote, which leads to a better standard.

Once a new international standard is identified as a requirement, then it is put on the work program of the relevant committee. Usually the new work item proposal contains some sort of basic draft showing what is required, along with a timetable. If there is a positive vote and at least four P members agree to take part in the work, the project is approved. The approved timetable outlines the time taken to complete the work and publish the standard. The same process is used for a revision of an existing standard, unless there is only a very minor change that can be done more simply. The directives for all IEC and ISO work can be found on their appropriate websites, and in most cases there is harmonization between the two bodies.

A working group is now set up to complete the drafting, which can typically take up to 3 years. The working group may, and often does, invite specialists from industry to help with the drafting. The finished draft from the working group now becomes known as a Committee Draft for Voting (CDV) and is sent out to all P members to vote and comment on for 5 months. If the result of the voting is noncontroversial and after any comments have been incorporated by the secretary of the committee, then the draft is recirculated as a Future Draft International Standard (FDIS) with another 2-month voting period. If the voting is positive, it is sent for publication. Problem cases might need to go back to the working group for resolution, but this is discouraged.

In the case of ship radio standards, it is likely that the radio performance must agree with recommendations of the ITU a UN specialized agency dealing with all aspects of telecommunications, including spectrum allocation and efficiency and noninterference. Recommendations of the ITU are usually normative references in international standards. ITU-R Study Group 5 currently deals with maritime mobile issues.

Once published, international standards have a shelf life. This can easily be extended if there are no changes needed and the management of the technical committee deals with this and any other issues at their biennial committee meeting, usually hosted by one of the P members.

Most P member and observer countries of ISO TC8 and IEC TC80 are also members of IMO; however, it might not be the same people or departments. In most cases, there is good cooperation. IMO is aware of the good work of these standards groups and again is prepared to reference international standards in *SOLAS* as a footnote, allowing easy changes in the future and being careful not to quote the edition number or the specific standard when there is a series, such as IEC 61162 series. This interface standard started as a simple interface and now includes fast Ethernet and other interconnection standards without altering *SOLAS* along the way.

5.1.1 Royal Majesty

When a working group starts on the road to an international testing standard against a particular IMO performance standard, it has a number of challenges. First, it should not require more facilities or performance than the original IMO standard. When drafting the IEC standard for GPS, the working group added receiver autonomous integrity monitoring (RAIM), a desirable performance monitor to improve safety. Some IMO member states were unhappy with this decision; however, after a passenger ship, the *Royal Majesty,* ran aground because of a faulty GPS antenna connection, the changes were accepted (see Figure 5.1). Second, it should ensure that any standards are ready in time to test and type approve new

Figure 5.1 Grounding of the ship Royal Majesty. [2].
(*Source:* National Transportation Safety Board).

equipment before any new IMO carriage requirement date. This really means that standards should be prepared in parallel with any IMO work on new performance standards. The *Royal Majesty* accident is an example of why agreed-on standards are important.

In 2012, the NI presented the findings of the accident and reported that "the grounding of the cruise ship *Royal Majesty* should be a warning to all navigators of the dangers of relying on one source of information. The ship went aground whilst on a cruise with over 1,000 passengers onboard. The GPS equipment had reverted to Dead Reckoning (DR) Mode after an antenna cable had parted and was thus not giving an accurate fix. In spite of being in coastal waters, the watch officers placed over-reliance on the information given to them by the automated features of the integrated bridge system [...].

About 52 minutes after the *Royal Majesty* left St. Georges, Bermuda, the GPS receiver antenna cable connection had separated enough that the GPS switched to dead reckoning mode, and the Autopilot, not programmed to detect the mode change and invalid status bits, no longer corrected for the effects of wind, current or sea conditions. Openly routing the GPS antenna cable in an area where someone occasionally walked, increased the risk of damage to the cable and related connectors" [1].

From a standards point of view, it was necessary to look at the protocols used by the GPS receiver to transmit information to the rest of the bridge navigation system and ensure that appropriate sentences would be available in the future to make other equipment aware that the GPS was not working correctly. The standard in use at the time was NMEA 0183, which was defined by the National Marine Electronics Association (NMEA) of the United States. These standards were adopted as the basis for an IEC standard IEC 61162-1 on interfaces. A modification to add appropriate sentences was quickly adopted by NMEA; however, because of the time scales at IEC, it took a further 2 years for the new standards to be published. The casualty of the *Royal Majesty* was high profile and most manufacturers adopted the changes before the international standard was available.

IMO performance standards are often a compromise—the time is short in the technical working groups of IMO subcommittees and members will often make a compromise to finish the work in the time frame. This could result in some misleading words in some standards, such as the "antenna should be substantially vertical." This means that later, when the international testing standard is developed, some extra work is needed to define what exactly this means. In the case above of A.694(17), the general requirements standard, two pages became an international testing standard of over 100 pages.

At its 82nd session, IMO's MSC recognized concerns that misunderstandings may exist among port state control officers regarding some of the type approval certificates issued for maritime products. The text as follows is taken from IMO MSC circular 1221 [2]:

> To clarify the purpose and procedures of the Type Approval certification scheme used by the flag States, classification societies and other recognized bodies, the Committee agreed that such scheme include the following steps; engineering evaluation; witnessing the manufacturing and testing processes; evaluating the manufacturing arrangements; and when all the forgoing procedures have been satisfactorily completed the issuing of a Type Approval Certificate generally valid for not more than 5 years which may be subject to annual inspections or verification of the manufacturer's process. The Committee has also agreed that the validity

> of the Type Approval Certificate itself has no influence on the operational validity of a product accepted and installed onboard a ship and that a product manufactured during the period of validity of the relevant Type Approval Certificate need not be renewed or replaced due to expiration of such Type Approval Certificate.

This means that equipment already fitted does not ever need to be updated unless the IMO specifically says so. As technology moves ahead, software-based products might no longer meet relevant international standards under this scheme and may cause interoperability problems.

For some equipment, this is now recognized and guidelines have been created to ensure that the latest software is used on certain equipment, especially ECDIS, where not being up to date is a safety concern.

5.2 e-Navigation Choices of Standards and Guidelines

One of the items identified early in the e-Navigation process was the need for a fast-track method for updating performance standards. Currently, any change proposed is considered as a new work item and therefore takes 5 years, as mentioned before. It also opens the standard to have changes made that may have been unwanted in the past version, an example being a country that did not use the common world geodetic system (WGS84) and wanted a control on all electronic chart systems to change the reference point for the chart. This was turned down on successive versions of the ECDIS performance standard.

When a particular item is identified in a performance standard as needing updating for a justifiable technical reason, that justification, including changed text, should be submitted to the IMO MSC, which should either agree on it at one session or send it to the appropriate subcommittee for one session and approve it at the next session of the committee. In this way references to other international documents, especially those of the ITU, can be updated quickly and easily.

For example, currently most DSC equipment is not compatible with the latest standards of the ITU and doesn't need to be under the existing type approval regime. The problem here is that DSC equipment on new ships with newer type approval may not be compatible with existing units on older ships.

If e-Navigation is to be adopted, a number of standards will need to be modified and without a fast-track process it could take years to see modified standards.

From time to time, it is necessary for IMO to clarify how equipment should work, be installed, or be used. On these occasions, the appropriate IMO committee or subcommittee drafts guidelines that are published as circulars, such as *Guidelines on Software Maintenance* and *Guidelines on Software Quality Assurance for e-Navigation*. They are therefore much easier to maintain, and in the early days of e-Navigation might be an easy way to provide standardization and harmonization without restricting innovation.

The e-Navigation gap analysis showed that a number of existing IMO performance standards were not entirely suitable for a fully integrated system of navigation and communications systems as envisaged in e-Navigation. With the current arrangements at IMO for revising standards taking 5 years and the ability of the NCSR subcommittee to handle only a few at a time, it was felt that a number of guidelines might be a way forward.

Three potential guidelines were initially identified as being important to ensure progress was made on a number of issues that had been identified during the gap analysis without having to revise individual equipment performance standards. These were *Guidelines on Human Centered Design (HCD) for e-Navigation*, used to ensure that any new equipment would be familiar and usable by the mariner, *Guidelines on Usability Testing, Evaluation and Assessment (U-TEA) for e-Navigation Systems*, so that designers and testers of such equipment could show that the equipment met the HCD requirements, and *Guidelines for Software Quality Assurance (SQA) in e-Navigation*, which ensured that designers of such software use proper software design and control techniques during development, which lead to properly supportable software.

During the further development of these guidelines, it was decided that there was a lot of common ground in the guidelines and so they were merged to become *Guideline on Software Quality Assurance and Human-Centered Design for e-Navigation* [3].

One other guideline was identified and developed ensuring that any test beds used during the development of e-Navigation were reported in a harmonized way, which is *Guidelines for the Harmonization of Test Beds Reporting* [4].

During the preparation of the e-Navigation SIP, the need for integration of equipment on a ship's bridge became clear. Some of the IMO performance standards already have interfaces and displays that might still be suitable for use in an e-Navigation context. However, some existing equipment performance standards do not currently have all the appropriate interfaces or use the appropriate up to date display standards, even though manufacturers might be providing them as an extra feature. For example, it is not currently an IMO requirement to be able to display AIS information on an ECDIS but some manufacturers' equipment has this facility.

Initially, a full list of all the relevant IMO performance standards that might need attention was presented to the NCSR Subcommittee [5]; however, with the use of guidelines the number of important standards was reduced considerably.

The SIP in particular noted that there was a need for a universal maritime data standard (UMDS). This standard was needed to harmonize the data structure used on board and ashore so that the data could be shared in a harmonized way. After some discussion, it was agreed that the IHO had such a data standard that was appropriate and could be used as the basis for UMDS.

The IHO S-100 standard provides a theoretical framework of components that are based on the ISO 19100 series of standards and specifications. These standards and specifications are also used as the basis for most contemporary geospatial standards development activities and are closely aligned with other standards development initiatives such as the Open Geospatial Consortium (OGC).

The IHO has also developed an associated registry that can be used in conjunction with the S-100. The IHO Registry contains the following additional components:

- Feature Concept Dictionary (FCD) Registers;
- Portrayal Registers;
- Registers of IHO producer codes.

The IHO Registry provides the infrastructure and mechanisms required to manage and maintain the resources listed above and to extend them as required. S-100 provides a schema and overarching management procedures for a registry and registers and the IHO Registry is implemented using these concepts.

IEC TC80 has adopted use of this standard data model and in time it will be incorporated within the existing IEC 61162 series of data interface standards.

5.2.1 Adding New Modules to the Revised Performance Standards for INSs (Resolution MSC.252 (83) Adoption of the Revised Performance Standards for Integrated Navigation Systems (INS))

The last revision of the IMO performance standards for INS before the e-Navigation strategy was approved made performance standards modular. "This means that provision for any new facility can be added to the performance standards by adding an appropriate module for that facility.

The modular concept of INS performance standards provides provisions for individual configurations and extensions where required. Currently, the performance standard contains four modules relating to integration of navigational information (A), operational requirement (B), alert management, (C) and documentation requirements (D) [...]

To reduce the risk of accidents that may result from important information not being acted upon due to, for example, lack of situational awareness or information overload, it is necessary to integrate received navigational information via communications equipment into the integrated navigation system in a harmonized and agreed-on way. In this way, information will be available at the appropriate display while not affecting the mandatory navigational displays.

e-Navigation relies on integration of relevant navigational information and INS provides an effective means to integrate navigation equipment data. By providing integrated and combined functions to avoid geographic, traffic, and environmental hazards, INS enhances the safety of navigation [6].

"Although module A of MSC.252(83) [7] is suitable for integrating navigation information required for e-Navigation, the INS performance standard will require two new modules so that information received by communications equipment can be integrated as well as properly displayed. The proposed two new modules relate to

- Harmonization of bridge design;
- Display of information.

A new module on the harmonization of bridge design will assist designers in realizing an ergonomic design of the bridge with the objective of improving the reliability and efficiency of navigation. This module will support the provisions of *SOLAS* Chapter V Regulation 15 relating to bridge design, design and arrangement of navigational systems, and equipment and bridge procedures" [6].

Another new module on display of information "will ensure that the INS can display information received via communications equipment. This module will outline the standardized interfaces for data exchange to support transfer of information from communication equipment to an INS interface so that information received via such equipment can be processed, filtered, routed, and displayed on the navigational system" [6].

New modules will "add functionality to the INS performance standard, which will facilitate a simplified and harmonized bridge design and ensure relevant information is displayed, including information received via communications equipment. This will result in reducing complexity without compromising existing navigational functionality in INS" [6]. This performance standard revision will mean that the related international standard IEC 61924-2 ed2 and its corrigendum will need attention. This refers to INS and any new modules will need to be tested during type approval, which means that this standard will need to be revised accordingly.

5.2.2 Draft Guidelines on Standardized Modes of Operation

Rather than changing all the performance standards of the onboard equipment to make them easier to use, a new guideline is suggested. As IMO states: "This will describe a standardized mode of operation and display for all navigational equipment and provide seafarers with the ability to operate all navigation equipment in a standardized manner, thereby improving the safety and efficiency of navigation.

Today with many different manufacturers of navigational equipment, displays and controls differ from one piece of equipment to another, which causes confusion for the mariner.

To rectify this problem, the IFSMA proposed and described on behalf of the NI "the concept of S-Mode for onboard navigation displays" in a submission to NAV 54 (NAV 54/13/1).

A challenge faced by all mariners is to become quickly familiar with the wide range of systems and models of navigational equipment supplied by different manufacturers [...].

Because of the increasing complexity and functionality of navigation equipment, a need has arisen for navigation systems to have more standardized functionality to enable better operation to support good decision-making. [...] S-mode calls for all navigation systems in the future to have a standard or S-Mode for display and control that when activated (with a single-operator action) defaults to a standard display (e.g., head-up display, relative vectors) and a standard user interface for key tasks.

The guidelines may also incorporate provisions for the configuration of personal settings that may be stored within the system and which would allow a user to rapidly customize the system to their preferred settings (overlay custom display features or give access to specialist information). [...] The benefit of S-Mode is that all shipboard navigation systems will have the ability to change to a standardized navigation functionality, by a single-operator action. S-Mode would supplement additional manufacturer-supplied modes. S-Mode supports the objectives of e-Navigation to improve navigation decision-making and hence safety and protection of the marine environment" [6]. If this guideline is approved it might still be necessary to look at the relevant testing standards to ensure that there is no conflict with the information advised.

5.2.3 Revision of the Guidelines and Criteria for Ship Reporting Systems (Resolution MSC.43(64))

Ship reporting systems and reporting requirements are used to provide, gather, or exchange information through radio reports. The information is used to provide data for many purposes, including SAR, VTSs, weather forecasting, and prevention of marine pollution. The existing guidelines and criteria for ship reporting systems (IMO Resolution MSC.43(64)) were initially adopted on December 9, 1994. This guideline outlined the criteria for planning, proposing, and implementing adopted ship reporting systems by contracting governments. In addition, general principles for ship reporting systems and ship reporting requirements are provided in IMO Resolution A.851(20). This further includes guidance for reporting incidents involving dangerous goods, harmful substances, and/or marine pollutants.

IHO states, "Currently there are no harmonized standards for ship reporting by electronic means and a considerable burden is placed on the ship to complete different paper forms for different identities ashore, such as customs, immigration, and cargo manifest. An automated electronic system will have the benefit of reducing administrative burdens. It is important, however, that the IMO ensures that a unified and harmonized system is adopted worldwide before regional systems are introduced.

In the absence of harmonized standards for ship reporting systems, national ship reporting systems may use different procedures and reporting formats. Such different procedures and reporting formats create an additional burden for ships moving from one area to another covered by different ship reporting systems. Such administrative burdens could be alleviated if ship reporting systems and reporting requirements were made in accordance with a single, standard format and procedures.

A revision and update to the existing guidelines will provide standardized and automated reporting of a ship's information through single entry of reportable information by electronic means. Single reporting arrangements are also be referred to as single window. The revised guideline will assist automated collection of internal ship data for reporting and automated or semiautomated digital distribution as required by coastal and port states. The required reportable information will contain both static documentation and dynamic information. The revised guideline will consider digital reporting formats based on recognized internationally harmonized standards such as the IMO FAL Forms (standard forms for reporting a ship's information on port arrival) or SN.1/Circ.289 (guidance on the use of AIS ASMs) as considered appropriate. The updated and improved guideline will support the provisions of *SOLAS* Chapter V Regulation 11 relating to ship reporting systems. The IMO updated this guideline in 2017.

The implications of updated guidelines will be that all countries wanting to take part in automated reporting will have clear guidance on how to implement this part of the e-Navigation strategy. Furthermore, it will be clear from the guidelines which standards will apply to the data exchange and data format. The benefit will be a worldwide harmonized automated ship reporting system saving cost and reducing administrative burden, while reducing the nonnavigational workload of the navigator and increasing the efficiency of trade" [8].

The development of any standards for ship reporting based on these guidelines will need the establishment of a CMDS based on the IHO S-100 series of international standards.

5.2.4 Revision of the General Requirements for Shipborne Radio Equipment Forming Part of the GMDSS and for Electronic Navigational Aids (Resolution A.684(17))

This IMO document states, "During the user needs analysis stage of the e-Navigation process, a large number of navigators advised that there was no indication of the quality and integrity of navigational data displayed [...]. Taking into account the safety of navigation, there is a compelling need to ensure that the navigator is presented with information that is accurate and reliable at all times with an indication when the equipment is not working satisfactorily.

Built-in Integrity Testing (BIIT) functionality will be a critical component of navigational equipment to ensure a level of confidence in their correct operation. It will provide the user with information about nonfunctionality of the equipment in an unambiguous and timely manner. [...] The BIIT will provide standardized self-check capability to ensure automatic quality and integrity verification testing for navigational equipment.

BIIT functionality will provide options for power-up testing, initiated testing, or periodical/continuous testing (in the background) of the navigational equipment. Such testing options may be provided by the vendor as part of the system application itself or they could be configured by the user during installation.

The type approval process for navigation equipment needs to be further developed to ensure that BIIT is included" [9].

5.2.5 Draft Guidelines for the Harmonized Display of Navigation Information Received via Communications Equipment

In the same document as above, IMO states, "Broadcast and reception of MSI by means of direct printing is an important part of the GMDSS. During the user needs analysis stage of e-Navigation, mariners expressed the need to sort and display MSI more effectively.

On most ships, MSI information received via communications equipment such as NAVTEX and INMARSAT-C are either displayed on separate screens or printed on a scroll of paper. The coordinates of the MSI must then be mentally compared to that of the vessel by the watchkeeper to assess relevance and risk. This is time consuming, distracting, and is susceptible to human error.

It is important that this information is displayed as task-oriented on the bridge and harmonized with other navigation-related information without obscuring critical navigation information. [...] A task-oriented integration and presentation of information, when all necessary information for the respective task and situation is available in a fast, reliable, consistent, and easily interpretable format will support the officers onboard in their decision-making and enhance the safety of navigation. [...] Although IEC TC80 has recently updated its test standards, IEC 62288 ed2, for the display of such information, it is necessary to review this work in relation to the reception of MSPs, which are defined below, for example.

It is important, however, that IMO ensures that a unified, harmonized, and user-friendly solution is integrated and adopted. The solution must be based on an agreed-on standard. [...] To fulfill the requirements for safe navigation to include all means and information in the decision-making, a presentation of this information at the navigational workstations is essential.

[...] The new guideline will include standard symbology and text support taking into account human element and ergonomic design principles to ensure useful presentation and prevent information overload. The guideline will consider IMO Performance Standards for the Presentation of Navigation-Related Information on Shipborne Navigational Displays (resolution MSC 191(79)).

This [guideline] does not introduce any significant additional burden to the maritime industry, but merely proposes that guidelines for the harmonized display of navigation information received via communications equipment are introduced. The proposal will have implications for equipment manufacturers and end users.

IEC 62288 ed 2, *Presentation of Navigation Related Information,* Annex A para 5.4, contains information related to the display of MSI that might be useful when further drafting the guidelines" [9].

5.2.6 MSPs

In IMO's document MSC/96/23/xx, further information is given, "As a result of user needs, gap analysis, and the IMO process leading to the development of e-Navigation, one of the prioritized solutions uses the principle of the MSPs [which are] digital messages about services provided by shore-based service providers to the ship for safe and efficient navigation. These messages can be created by many different service providers and may contain free text as well as position information (LAT/LONG) and other numeric data.

[...] This concept contains relevant information provided by shore-based service providers to the ship for safe and efficient navigation. Several initiatives are already ongoing in regional projects, and in order for a global solution to work, an IMO decision such as a resolution is needed to coordinate and harmonize the format of these services provided by shore-based and other stakeholders for delivery to ships.

[...] This output should ensure that the development of many different MSPs, which in some cases are developed by other organizations, such as IHO and IALA, are done so in a coordinated and harmonized manner globally.

[...] The delivery is the responsibility of the service providers. The harmonized communication and display on board and ashore of MSPs requires an agreed-on technical structure and format.

In the long term and for complete harmonization of MSPs and their display on board, it may be necessary to revise the requirements contained in *SOLAS* Chapter IV and V and some of the supporting performance standards for the relevant equipment" [10].

As the IMO has ownership of the development of e-Navigation, it should continue to lead and coordinate future work on e-nNavigation in order to ensure global implementation in a structured and coordinated manner (IMO document NCSR1/28 para 9.5). The e-Navigation SIP facilitates coordination of efforts by relevant subcommittees, related international organizations, member states, relevant regional bodies, and the maritime industry.

"The lack of harmonization could cause many different systems and protocols to be adopted, leading to interoperability issues, and may cause problems with the display of the information on board ships and ashore. Therefore, there is a compelling need for harmonization" [11] including synchronization of the functional and physical links used for exchange of information as described in IMO document NCSR 1/28 Annex 7, (Figure 5.2), and a description of the ship and shore architecture for prioritized solutions.

As mentioned in Chapter 3, the process description and information exchange interfaces, together with user needs, were the basis for the specification of electronic services provided from ashore to ships. This also represented an important outline of the development of technical architecture that assisted in achieving the objective of e-Navigation. The architecture safeguarded that work is conducted in a systematic manner and that continuous processing of information was kept as the focus of the work.

Figure 5.2 shows the principle of an information/data flow in the e-Navigation architecture. The figure shows the complete overarching e-Navigation architecture and defines two additional important features:

1. CMDS, which spans the whole of the horizontal axis;
2. The World-Wide Radio Navigation System (WWRNS).

In addition to identifying the concept of MSPs, the architecture also brings into focus the operational service level, the functional links used by technical services, and the physical links used by technical services. Furthermore, the architecture highlights the fundamental distinction between information and data domains, explaining the relationship between user-requested information items and introducing the concepts of operational and technical services, as well as functional and physical links, into a hierarchical

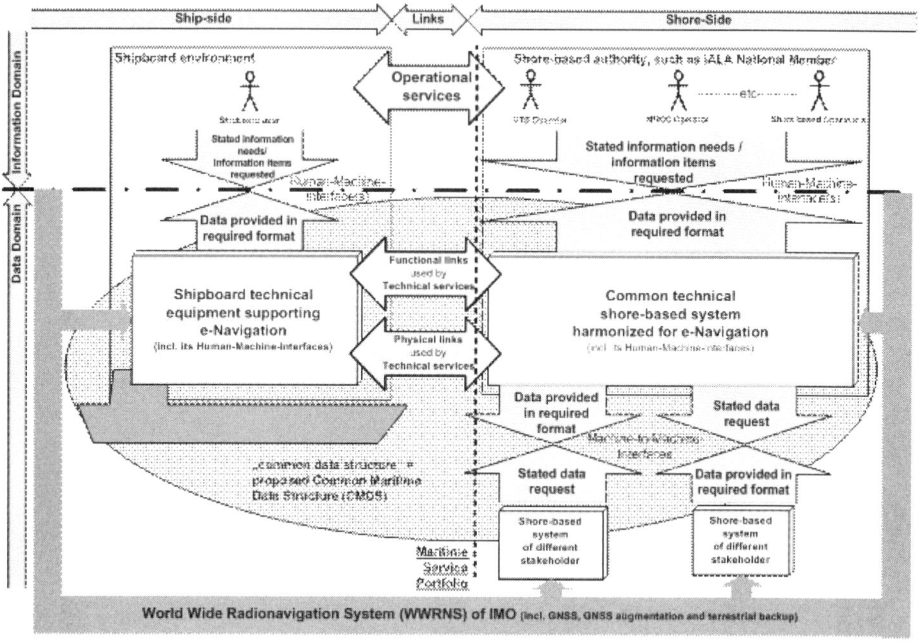

Figure 5.2 Overarching e-Navigation architecture: identification of communication systems for e-Navigation. (Source: NCSR 1/28 Annex 7.)

perspective. It also unfolds the relationship of shore-to-shore data exchange.

"In order to provide a harmonized solution, it is envisaged that all data delivered to a ship should be processed in such a way that the data is organized and delivered to the appropriate equipment" [11].

In creating an e-Navigation architecture, it is important to identify information and data flows and the interaction between applications and user interfaces. Consequently, there needs to be a data structure to optimize the use, interoperability, flow, and accessibility of relevant information and data within the maritime domain, including both ship and shore aspects. It is therefore important to harmonize efforts in data modeling with the aim of creating and maintaining a robust and extendable maritime data structure.

In order that the information is displayed on board in a structured way, it is necessary to agree on formats for the layout of such information so that all providers can be clear about how to structure their messages.

Some information containing position data will allow MSPs to be displayed on other equipment, such as ECDIS and radar where appropriate, and in order for this to be successful, the position information must be in the correct format so that it is machine-readable.

The benefits of this structure are internationally harmonized and agreed-on functional and physical links used for the exchange of information required for MSPs, as shown in IMO document NCSR 1/28 Annex 7, (Figure 5.2).

No overarching international standards currently exist for harmonizing the functional and physical links used for the exchange of information required for the MSPs. Some relevant international organizations (NGOs/IGOs) are already developing such standards in support of some MSPs. This work is progressing in close cooperation with other organizations. Both IALA and IHO are developing suitable guidelines for relevant MSPs that will support this initiative.

5.3 Carriage Requirements for e-Navigation

Currently, the IMO does not intend to mandate the carriage of e-Navigation equipment on board. This makes the standardization of equipment much easier because type approval is usually related to a carriage requirement. However, any equipment fitted to a ship's bridge will have to be tested for interface capability (see the *Royal Majesty* accident detailed above) and environment and EMC issues as required in the IMO general requirements performance standard A.694(17). As the concept of e-Navigation is introduced, it will become necessary to reevaluate the standards being used, and if older ships with out-of-date but approved equipment on board will be able to take part. It is likely that a number of bigger ports will require some e-Navigation features to be operational on a visiting ship. This could include the automated transfer of pre-arrival information and the ability to receive and display MSPs, especially in a port where digital information is the norm. This may cause the shipowner to update his or her equipment, and at this point, certification may become necessary. Certainly, a better clarification of what constitutes type approval is needed and MSC circular 1221 will need a radical rethink.

Some flag states today do not have national authorities to certify ships and use recognized organizations (ROs) to do so on their behalf. These organizations are sometimes strict about equipment installation—limiting the installation to the minimum IMO requirements, thereby minimizing the insurance liability of using nonapproved equipment. It would therefore be difficult to add e-Navigation features in these cases without some extra standardization. It is obvious that the performance standard on general requirements A.694(17) and the relevant IEC testing standard IEC 60945 should be used to ensure reliability and compatibility with existing equipment.

References

[1] The Nautical Institute, *Official Report: Grounding of Cruise Ship*, available at http://www.nautinst.org/en/forums/mars/mars-2012.cfm/cruiseshipgrounding, accessed May 10, 2017.

[2] IMO, *Validity of Type Approval Certification for Marine Products*, Document MSC.1/Circ.1221, 11 December 2006, available at https://www.iho.int/mtg_docs/industry/ECDIS_workshop_12/MSC_Circ1221.pdf, accessed May 10, 2017.

[3] IMO, *Guideline on Software Quality Assurance and Human-Centred Design for e-Navigation*, Document MSC.1/Circ.1512, 13 July 2015, available at http://www.ics.org.ir/Downloads/CLD/News/MSC.1-Circ.1512%20-%20Guideline%20On%20Software%20Quality%20Assurance%20And%20Human-Centred%20Design%20For%20e-Navigation%20(Secretariat).pdf, accessed May 10, 2017.

[4] IMO, *Guidelines for the Harmonization of Test Beds Reporting*, Document MSC.1/Circ.1494, 21 November 2014, available at https://www.iho.int/mtg_docs/enc/e-nav_documents/English/MSC_Circ1494-Guidelines_On_Harmonization_of_Testbed_Reporting.pdf, accessed May 10, 2017.

[5] NCSR1 INF.6.

[6] IHO, *Proposal on Drafting New Modules to Performance Standards for Integrated Navigation System (INS)* (Resolution MSC.252(83)), Document WWNWS8/3/4/1c, July 26, 2016, available at https://www.iho.int/mtg_docs/com_wg/CPRNW/WWNWS8/WWNWS8-3-1-2c-Proposal%20on%20drafting%20new%20modules%20to%20Performance%20standards%20for%20integrated%20navigation%20system.pdf, accessed May 10, 2017.

[7] IMO, *Adoption of the Revised Performance Standards for Integrated Navigation Systems (Ins) Annex 30*, Resolution MSC.252 (83) adopted on 8 October 2007 [online]. Available: http://www.imo.org/en/KnowledgeCentre/IndexofIMOResolutions/Maritime-Safety-Committee-(MSC)/Documents/MSC.252(83).pdf, accessed May 10, 2017.

[8] IHO, *Circular Letter 20/2015*, 6 March 2015 [online]. Available: https://www.iho.int/mtg_docs/circular_letters/english/2015/Cl20e.pdf, accessed May 10, 2017.

[9] IMO, *Work Programme*, Document MSC 95/19/xx, 3 March 2015, available at http://www.ics-shipping.org/docs/default-source/Submissions/IMO/work-programme---implementing-e-Navigation-to-enhance-the-safety-of-navigation-and-protection-of-the-marine-environment.pdf?sfvrsn=2, accessed May 10, 2017.

[10] IMO, *Work Programme–e-Navigation*, Document MSC 96/23/xx, February 2016, available at https://www.iho.int/mtg_docs/com_wg/NIPWG/NIPWG2/NIPWG%202-33.1%20%20MSC96-new%20input%206%20(MSPs)-v1.pdf, accessed May 10, 2017.

6

The Future

6.1 Introduction to the Future

In his book, *Grownup Digital: How the Net Generation is Changing Your World* from 2008, the Canadian author Don Tapscott described how at that time young people were at the cusp of a vast change in social development. "They've grown up with the internet," he observed, "and digital media is as natural to them as breathing air. This is the first-ever global generation, and in the US they have just elected their [own] first president [Barack Obama]" [1].

The baby boom generation, Tapscott argued, passively watched TV. Today's net generation is a more active participant, engaging in media multitasking and social networking. Although online gaming and other forms of entertainment remain hugely popular among this demographic, Tapscott pointed out that a sense of shared scrutiny and self-organization is becoming more prevalent. User-driven online communities such as Facebook, MySpace, YouTube, and Flickr demonstrate just how many people at that time—especially the young—have "grown up digital," observed Tapscott.

"We are growing increasingly accustomed to functioning in self-organised communities and self-organised programmes. We operate with degrees of freedom, speed and innovation unknown by previous generations," he pointed out. "Social networks and wiki cultures draw together like-minded people from around the world, presenting them with a tool or a platform that supports their self-defined goals and aspirations" [1].

Ten years after, thousands of these young people have educated themselves in the use of technology in addition to their formal education and now work in the maritime sector, on board ships, or at shipping companies, at ports or VTSs, maritime universities and research institutes, insurance and financial institutions, maritime administrations, or in the maritime manufacturing industry. When they do their "net" working they take for granted that systems are harmonized, integrated, and standardized. This is one of the key areas of e-Navigation development. They expect that the net platform works satisfactorily and that they have easy and rapid access in a secured manner to whatever they want to look at. While their lives revolve around computers and cell phones that work in an integrated way at home, their business lives in the maritime sector are separated by a lack of harmonization and interoperability of the existing systems and a lack of good and robust connectivity with ships and ship-to-shore systems. This is due to a lack, over many years, of addressing the problems at an international level for connectivity of maritime systems. e-Navigation is a way of solving this problem.

6.2 Digital Globalization

In 2016, the McKinsey Global Institute (MGI) published the report *Digital Globalization: The New Era of Global Flows* and in the report, their analysis found that global flows of all types support growth by raising productivity, and data flows are amplifying this effect by broadening participation and creating more efficient markets [2].

The MGI Connectedness Index ranks 139 countries by taking a comprehensive look at how countries participate in inflows and outflows of goods, services, finance, people, and data. This also includes testing the impact of all types of flow together and considering how countries are positioned in each web of flows. The index found large gaps between a handful of leading countries and the rest of the world. Singapore topped the rankings, followed by the Netherlands, the United States, Germany, and Ireland. Number six in the rankings is the United Kingdom. China, which has grown more connected, is number seven on the list.

According to the report, countries that participate in global flows can specialize in what they produce, realize economies of scale, and invite competition, which spurs both efficiency and innovation in local industries. These benefits have been recognized since they were outlined by economist David Ricardo in the early 19th century—but today there is a difference. As they transform traditional flows and enable broader participation, digital technologies are amplifying these effects.

Global flows of data primarily consist of information, communications, transactions, and intracompany traffic. Container ships still move products to markets around the world, but now customers order them online, track their movements, and pay them via digital transactions [2].

Looking at the countries at the top of the list, they all represent a large maritime sector and they all have a relation to e-Navigation activities. The first four countries on the rankings domiciled some of the authors of the paper submitted to the IMO regarding the proposal of having e-Navigation on the IMO agenda through the development of an e-Navigation strategy, as described in Chapter 1. The submitters believed that it was appropriate that IMO developed a broad strategic vision for incorporating the use of new technologies in a structured way, ensuring that their use is compliant with the various electronic navigational and communication technologies and services that are already available.

Singapore, which is at the top of the list, has been an active participant in e-Navigation development for many years. As a global shipping hub it looks like its approach to e-Navigation has been firmly anchored with a view to attract maritime business by offering new, efficient, and simplified digital solutions for ships when using the services of the Port of Singapore. At the same time, it has contributed to increasing the safety level for ships sailing in its waters. Its focus of attention has especially been on the means for standardized and automated ship reporting in a single window system and the e-Navigation solution for shore-based digital services provided to ships (MSPs). More of its e-Navigation work is described in Chapters 1 and 4. According to MGI's report [2], the globalization journey of Singapore began decades ago when it merged as Southeast Asia's global shipping hub, particularly for oil and fuel.

Number two on the list, the Netherlands, contributed to regulatory issues with focus on the enhancement and development of future services between shore-based authorities and shipping, their integration within the existing system, as well on the development of communication, data, and PNT systems contributing to safe and efficient maritime transport in coastal waters, harbor approaches, and in ports.

The United States, number three in the rankings, has been through participation at IMO, IALA, IHO, and various other bodies, and a supporter and contributor to e-Navigation development. Its e-Nav SAP has established a vision for U.S. e-Navigation implementation, which is described in Chapter 4.

Germany, number four on the list, has worked in particular with regulatory e-Navigation issues, focusing on integration and presentation of available information on navigational displays onboard received via communication equipment. Ireland, number five, has carried out work on The Dublin Bay Digital Diamond (DBDD), which is an e-Navigation demonstrator project for the Dublin Bay area, the purpose of which was to provide an opportunity for users across the maritime sector to explore the potential of e-Navigation services. The primary objective of the project was to demonstrate and develop e-Navigation services that improve the safety and efficiency of maritime transport.

Number six in the rankings, the United Kingdom, originally introduced the e-Navigation concept and has during the IMO development been very involved in the work. China, next in the rankings, has, as described in Chapter 4, worked on the further development of e-Navigation solutions including being the organizer of e-Navigation testbeds; an interesting project completed in the Port of Shenzhen "made the distinction between dynamic information, which is essential for navigational safety, and business-related information. The latter type of information should not unduly burden ship bridge personnel, and should preferably be transmitted using standard electronic tools such as the Internet, e-mail, and electronic data interchange (EDI). China suggested that the test showed that AIS could exchange most of the dynamic information messaging" [3].

We could continue with presentations of other countries on the list and how they contribute to e-Navigation work; however, all the countries involved have an important goal in common; namely, conducting maritime national and international business activities, while at the same time demonstrating a positive attitude and a strong will for introducing digital means. The potential is huge: according to the ICS, "around 90% of world trade is carried by the international shipping industry. Without shipping, the import and export of goods on the scale necessary for the modern world would not be possible. [...] There are over 50,000 merchant ships trading internationally, transporting every kind of cargo. The world fleet is registered in over 150 nations and manned by over a million seafarers of virtually every nationality" [4]. Global or regional port statistics are difficult to ascertain because there is no global organization responsible for collecting this data, but the United Nations Conference on Trade and Development (UNCTAD) reports that there are thousands of ports worldwide; some estimates put the figure at over 10,000 [4].

6.3 Challenges

The maritime industry is facing more challenges than ever to business-as-usual conditions, from technology changes—mostly from the digital sector—and from environmental regulations. Writing in *Container Shipping and Trade*, Martyn Wingrove saif, "outside influences could threaten the current status quo in shipping and provide the next generation of solutions. For example, companies such as Amazon, Google, and Huawei Technologies are looking to deliver new digital technologies in the shipping sector, which could revolutionize the industry. They could become digital disruptors to the conservative shipping community by offering different business models and technologies" [5].

According to the Marine Electronics and Communications report, during the Disruptive Sustainability discussion between different thought leaders, held in the Baltic Exchange in London at the end of February, 2016, executive vice president of Global Marine at the American Bureau of Shipping, "Kirsi Tikka, said the shipping industry needed to create the future, not be dragged toward it by these outside influences. It should be investing in better analytics and data processing and in more shore-side support, and it should be innovating to solve the major issues. 'We need to open up to a wider pool of ideas and be open minded to outside people,' she said. 'We need to think differently and respond rapidly, and we should attract innovation'" [6].

She thinks that "classification societies can be an incubator or a platform for innovation. 'We are now looking for more disruptive innovation from automation and big data analytics. But for the most part, we do not see a road map of how we get there,' Dr. Tikka continued. 'This is mainly because it involves expertise not found in our traditional industry, and leadership that understands how to apply and benefit from the technology.' Shipping needs an innovation strategy to tackle the operational issues and regulatory challenges and find ways of being more competitive and environmentally and financially sustainable. The industry needs to harness the potential from large-volume data processing and analytics and fast communications between ship and shore.

'Smart use of data means we can be less intrusive and more predictive in the way we work with asset owners and companies, and be less bound by calendar-driven events,' said Dr. Tikka. 'But we need to be confident about the quality and the reliability of the dat'" [5].

Avoiding accidents and ensuring safety of onboard personnel and passengers presents one of the most complex challenges faced by the shipping industry as, unlike mechanical or technical systems, safety systems must account for the seemingly infinite variables of human behavior. Operational safety also relies on a broad range of technologies. Satellite and communication technology and surveillance and navigation technology have significantly changed the way ships are navigated and the planning of routes. Those technologies also enable improvements with regard to distress and safety systems as well as SAR activities. An increasing need for data transfer between vessels and onshore bases to ensure optimal operational efficiency, safety, and security are driving these developments. Additionally, issues such as interoperability between different systems and the capacity of the systems and services become more important.

Interactions between technological systems and the human element are becoming more complex and require a better understanding at the design, manufacturing, and operational phases of the product and system development. The development of global standards, principles, and guidelines, such as IMO's e-Navigation concept covering ergonomics and improved integration of bridge controls, is already underway, helping to ensure better coordination between design and operations. In addition, a more standardized bridge in line with e-Navigation will reduce the time needed for personnel to familiarize themselves with bridge controls and eventually lead to development of best practices that can be applied throughout the industry. Harmonized modes of operation will play a key role in this respect.

According to the IMO, the main challenges for the maritime community are to increase and improve the utilization of these new navigation and communication technologies through their implementation and providing training to seafarers. Another challenge will be to update relevant safety procedures in order to keep them up to date with current technology developments. This includes update of the GMDSS as well as harmonization of international SAR efforts. Additionally, the IMO will have to support the education and training of seafarers to increase their awareness of safety and security issues of electronic navigation, as well as to understand the dangers of overreliance on electronic navigation systems [7].

6.4 Ships, Ports, and VTSs in the Future

The implementation of e-Navigation is an innovative process as shipping moves into the digital world and will provide digital information and infrastructure. So, what will be the near future of shipping? Views on the future are presented by DNV-GL in its report, *Technology Outlook 2025.* The report says that "before 2025, many ships, systems, and components will be linked to the Internet, making them accessible from almost any location.

Modern ships are becoming highly automated and are increasingly dependent on software-based control systems. Advanced software and simulation capabilities will result in more advanced systems being controlled by software, while near real-time evaluation possibilities will be available, accompanied by suggestions for corrective actions by the crew and providing supply chain management decision support" [8].

New technologies on the horizon for the coming 10 years include extension of cellular networks in coastal areas, VDES (the new data service on the VHF band incorporating AIS), Wi-Fi in ports, and, most important, satellite communications, improving coverage and bandwidth. Currently, the maritime industry contributes to the growth in deployment of VSAT equipment on board ships. Most classed vessels will be broadband capable in the future. As well, the VSAT network capacity is increasing owing to the introduction of new HTS systems, with 2 to 10 times higher throughput than classical satellites.

Most ships, systems, and components will be linked to the Internet, making them accessible from almost any location. At the same time, combining data streams from multiple sources will boost performance management (including fleet utilization, routing, trim, fuel consumption, and emission management) and asset integrity management, building on remote condition monitoring as well as allowing for an increased level of automation. This may, in turn, facilitate remote-controlled and autonomous ship operations.

Improvements in maritime connectivity will allow supply chains to be more efficiently organized around adaptable operations. This could reduce lead times and fuel consumption by optimizing arrival times and allow a better organization of operations and workforces on land for handling cargo and carrying out repairs and inspection.

Apart from enhancing safety and efficiency, ship connectivity will also answer the need for more transparent operations and help build trust and collaboration between various industry stakeholders based on the collection and analysis of shared information. Ship connectivity will provide a unique opportunity for maritime authorities to monitor compliance with existing regulations to improve safety, achieve environmental targets, and boost competition in the industry, according to the report [8].

This overall future description could have been a likely e-Navigation scenario, and as previous views expressed in this chapter, the IMO's chosen direction of the implementation of e-Navigation based on e-Navigation solutions correspond with the thoughts provided in the DNV-GL report. However, it looks like more stakeholders in the future will be involved in technology development and implementation, including those from outside the traditional industry. Therefore, instead of considering outside influences as a threat to the current status quo in shipping, they should be invited to contribute to a common innovation strategy in shipping, which must also tackle operational issues and regulatory and environmental challenges. IMO's e-Navigation implementation strategy should be a basic part of this, as its gap analysis showed that shipping needs harmonization of equipment and systems more than ever.

The International Association of Independent Tanker Owners (INTERTANKO) is the association of the owners of independent tankers in the world, and strives to achieve the goals of zero fatalities, zero pollution, and zero detentions.

Senior Manager Marine, Captain Johan Gahnström said that "INTERTANKO is a strong supporter of e-Navigation" and the organization "would like the development to be focused on empowering the officers onboard with more information for an enhanced situational awareness. [However,] INTERTANKO do[es] not endorse development where the shore authorities aim at commanding the vessel. INTERTANKO welcome a shared situational awareness between ship and shore" [9]. He emphasized that there should be focus on two important things when taking e-Navigation to the next step:

- Zero groundings;
- Zero collisions.

He recommended "thinking of the benefits for navigator and ship safety when designing next generation of tools/services. All information available, always, may not be the answer," he said. Captain Gahnström provided an example: "The variations on ECDIS displays today are not acceptable. Moving forward to e-Navigation, [he] would like to see a change and more commonality between systems" [9].

He also reflected on how e-Navigation will affect the members of INTERTANKO: "We will be able to have a variety of information readily available on the bridge transmitted from the shore side. This may need additional equipment and systems onboard tankers, and there will or may be a cost involved. If so, e-Navigation has to prove that it will significantly contribute to achieve the goals of zero groundings and collisions," the Captain ascertained [9].

6.4.1 Future VTS

The complexity of utilization of the seas is growing, threatening the maneuverable space for shipping. The need for proactive management of vessel traffic in these areas is thus likely to increase, further driving the interaction between ships and shore authorities. Management of operational space from a shipping perspective by evolving VTS, supported by the capabilities of e-Navigation and its MSP developments, and in conjunction with Marine Spatial Planning are seen as candidate combinations on how to deal with the challenges for safe, secure, and efficient navigation in clean waters.

One of the aims of e-Navigation is to reduce the misunderstanding between the OOW in the VTS center and the bridge team, avoiding information overload while ensuring that the information is delivered in a context-sensitive manner.

A future consequence of the evolution of VTS is the focus increasingly being placed on ensuring the general efficiency of vessel traffic from both operational and commercial perspectives. VTS is increasingly being utilized as a means to optimize the flow of vessel traffic to provide benefits to ship operators and ports ranging from enhancing fuel efficiency through to managing vessel traffic to facilitate optimal berthing windows and time alongside.

Another aspect of the constraints of navigation is that traffic flow is increasingly squeezed in between oil or gas production, wind and fish farms, and so on, reducing the available sea room for safe navigation. Consequently, ships need more information, more assistance and more services, which is by essence the role and responsibility of VTSs.

Adding to this pressure, tidal windows and lock planning sometimes must be taken into account. An efficient sharing of relevant information is therefore essential. The information must be shared in a timely manner between all relevant stakeholders in an effective, secure, and trustworthy manner, ensuring cybersecurity issues, including validity and confidentiality of data.

"The role and expectation of VTS is expanding rapidly. AIS has changed the maritime World. There is no more secrecy in the movement of ships. This places a new responsibility on the shoulders of Authorities, including Harbour Masters. It is not possible any more to say 'I did not know.' If you do not know it is because you have not deployed the necessary equipment with the competent staff. And if you know, you need to take the necessary actions to make the traffic safe, secure, green and efficient. This is a large part of the role of VTS, including tomorrow to take into account the ships navigating unmanned or with a very limited staff on board," [10] wrote Francis Zachariae, secretary general of IALA, in a paper to the IHMA Conference. He stated: "Digital connectivity is the key; remember that this is a two way fairway, the navigator needs information from the harbour and the harbour needs information from the navigator" [10].

6.4.2 The Future Port

According to the IMO, the operational complexities of dealing with large ships, the need to manage congestion, the need to do more with less space and the continued pressure to enhance safety and security in and around port areas and to embrace greener technologies and working practices, are among the specific developments that the entire supply chain needs to address. "Port infrastructure, support facilities, and the human capital throughout the maritime industry need to be further developed to enable ports to offer a more efficient service," the IMO document, *Trends, Developments and Challenges Facing the IMO in the 2018–2023 Period,* stated [11].

Vice president Lene Vesterlund at Kongsberg Norcontrol AS, a big supplier worldwide of equipment and system to ports and VTSs, believed efficiency processes would continue for ports in the future in order to remain competitive as well as sustainable. "Increased efficiency can be achieved through Just-In-Time (JIT) arrival, which leads to less time spent waiting for resources such as pilots, tugs and berths, and onto terminal resources such as cranes and trucks. Just-In-Time arrival is possible through the integration of all relevant port systems, including vessel traffic management systems, information management systems, scheduling systems used by allied service operators (pilots, tugs, etc.), terminal operator systems, and the wide range of transportation services that might be available, depending on the port and port systems. Port planning related to the arrival and departure of vessels will be increasingly connected to the entire transport logistic chain," she said.

From this port and VTS supplier point of view, automated operations and digitization will simplify the navigators' administrative burden, ensuring that relevant and accurate information is sent accordingly. The likely result of automation and digitization will be increased safety and reduced risk both at sea and in port; navigators will be able to focus on navigation as their administrative burden is reduced. One example of this would be automated ship-to-shore reporting. More services will be provided from shore to ship, leading to new revenue streams for the ports and/or the vessel traffic centers; for example, information service such as weather data, traffic/arrival data, and routing services.

In addition, autonomy is quickly entering the maritime and port domain with autonomous vessels at sea, as well as autonomous cranes and trucks on shore, and small boats near shore. Future ports will need to be ready to handle the arrival of an autonomous vessel and will also need to decide if/how autonomous vessels can be used to deliver services.

However, with a rise in digitization, security threats from cyberattacks become ever more relevant. Future ports will have an increased focus on security as a whole, and on cyber security specifically, in order to guarantee that the data and information the ports own is protected.

New security technologies, such as drone surveillance, will be available and more widely used. Underwater threat detection could also become increasingly attractive, especially in regions with higher-than-average levels of threat due to conflict and/or terrorist activities.

Safety will remain critically important for future ports as they continue to seek ways to reduce pollution, including oil spills, as well as dangerous and unwanted navigational situations, both of which can endanger port operations and negatively impact the environment. Behavioral analysis technology will be critical for larger ports as a means to improve decision support to shore-based operators, which will increase the safety of vessels in port.

"As the technology needs of ports will vary from location to location, technology systems will need to be adaptive. Some ports will seek to consolidate and integrate operations; other ports will opt for distributed operations centres. Sound levels will be an issue in large, integrated operation centres, so ergonomics will be important. Integrated operations centres will require integrated technology solutions that incorporate customizable workstations with latest display technology, such as touch screens, large high-resolution monitors, and even smart glass. For distributed operations centres, integration with portable devices, wearable technology, and cloud based solutions will be of greater interest," Lene Vesterlund said.

Captain Kevin Richardson is president of the IHMA. He said that "e-Navigation is still in its infancy,... but will evolve fast. Ports need to anticipate its implications and grasp the opportunities it affords. The maritime industry is traditionally slow to embrace new technology; ... this needs to change" [12]. He argues that communications will become more standardized, and data exchange between ports and vessels will become more complex and more regular....

"Ports and the key people in ports need to prepare and invest for the future e-Navigation systems which should support good decision making, enhance performance and reduce errors. VTS' and terminal operations need to support people and the key decisions they need to make such as analysis, ranked options, key information relevant to the situation and risk levels, and controls for staff to make the right call at the right time to deal with safety, security or environmental incidents. Ports should invest in e-Navigation systems, plant and equipment if they want their business to grow, prosper and avoid costly delays, inefficiency and in the worst case serious or catastrophic accidents," [12] Captain Richardson emphasizes.

6.5 Moving e-Navigation Onboard and Ashore

An interesting question about e-Navigation in the future process is how do we do it on both the ship and the shore now that all the pieces are on the table? The first issue could be how could we convert to e-Navigation equipment and systems on board?

When it comes to e-Navigation solutions regarding the workable and practical use of the information and data on board, e-Navigation solution (S1) on improved harmonized and user-friendly bridge design and the solution (S4) on integration and presentation of available information in graphical displays received via communications equipment are possibly related to an INS display. There are to be additional modules for INS such as situation display and user-friendly design, as described in Chapter 5. Like the introduction of ECDIS, e-Navigation solutions have already—as presented in Chapter 3—been through IMOs FSA for risk reduction and cost benefit.

With regard to standardized modes of operation that are a part of the first solution, S-Mode (S1) is new equipment that will have implications for shipbuilders, classification societies, competent authorities, end users, and the equipment manufacturers. The benefit of S-Mode is that all shipboard navigation systems will have the ability to change to a standardized navigation functionally.

However, in order to support the links between solutions S1—user-friendly bridge design and S-Mode—and S4—information in graphical displays—and prevent cyberattacks, the guideline on SQA and HCD should be mandatory in the long term. The IMO states that "Navigation systems increasingly provide a variety of information and services for enhancing navigation and efficiency. These systems require the connection and integration of onboard navigational systems as well as shore-side support systems and involve the collection, integration, exchange, presentation, and analysis of marine data and information. The merits of navigation systems can be found not only in their range of functions but also underpinned by their trustworthy software and overall usability. Achieving trustworthy and usability in the development of complex systems requires a disciplined and structured approach. The guideline encourages such an approach in the development and management of e-Navigation systems" [13]. It also supports the principal requirements specified under SOLAS regulation V/15.

The solution (S3) on improved reliability, resilience, and integrity of bridge equipment and navigation information will become mandatory when IMO resolution A.694(17) (General Requirements for Shipborne Radio Equipment Forming Part of the GMDSS and for Electronic Navigational Aids) is modified to add BIIT. It should also be considered an INS functionality as well as being related to the guideline on SQA and HCD.

When it comes to efficient transfer of marine information and data between the users, such as ship to ship, ship to shore, shore to ship and shore to shore, there are some aspects that are very interesting.

In order to establish e-Navigation solution (S2), on means for standardized and automated reporting, existing ships might need new equipment on board that could be needed for:

- Links to existing communications equipment;
- Links to the navigation system GPS, AIS, and ideally an INS where all the data is integrated;
- Contains software that collects the information where possible automatically, arranges transmission securely, and takes care of handshaking to ensure reliable links using existing standards.

It also might need a server ashore with interfaces that:

- Enables reception from ship with handshake;
- Enables communication with other shore-based stakeholders;
- Ensures security of the information.

An automated electronic system for ship reporting will have several benefits for both the ship and shore sides and will reduce administrative burdens. Ship reporting systems are linked to the SOLAS and FAL Conventions.

It is important, however, that IMO ensures that a unified and harmonized system is adopted worldwide by regulations before regional systems are introduced. As described in Chapter 2, today's ship reporting systems are challenging, and possible solutions are mentioned in Chapter 4. Ashok Mahapatra, IMO director, said that "ship reporting is one of the main issues in the IMO's consideration of e-Navigation [...]. In the absence of harmonized standards for ship reporting systems, national ship reporting systems may use different procedures and reporting formats, which create an additional burden for ships moving from one area to another. Such administrative burdens could be alleviated if ship reporting systems and reporting requirements were made in accordance with a single, standard format and procedures" [9].

What about improved shore-based services provided to ships, e-Navigation solution (S5) on MSPs? As described in Chapter 4, a lot of development work is going on regarding MSPs, which include improved communication of shore-based service portfolios. There is therefore, on the ship side, navigation information received from communications equipment; that is, e-Navigation solution S4, information in graphical displays.

Service providers of maritime services need to have access to suitable communications channels and be ready to transmit MSPs in a standardized and harmonized manner to all ships globally. This may require coastal states to develop systems for providing electronic information, taking into account the future revision of the GMDSS and new systems such as VDES. However, the MSP guidance should not define the detailed content of a particular MSP or aim at harmonizing the service itself. This is the responsibility of the relevant data and service provider.

Improved satellite connectivity and digital access to information allows ships to be online almost all the time. By satellite, connection data is shared and jointly used by the crew on the ship, the owner's operational centers onshore, actual ports, and VTSs. This allows operating systems more safety with greater efficiency and less environmental impact. This integrated way of operating connects the key stakeholders onboard and ashore and reduces the chance of information getting lost, helps identify where the risks and inefficiencies are, and allows making decision quicker. The technical architecture is important, as well as interoperability, which is the property of a product or service whose interfaces are completely understood. In the e-Navigation world, this means that any ship can work with another ship or the shore because they are using the same interfaces.

"The human element is still vital—it is simply applied in new ways. e-Navigation is a good example of the very human way of dealing with change. The real challenge is expanding our mindset so that ICT is applied on much bigger, global scale—and come to think of ICT not only as a new, extra tool, but as the foundation for the way in which business is done if competitiveness is to be retained," Andreas Nordslett, director-general of Danish Maritime Authority, said at e-Navigation Underway in 2016 [15]. He raises a central issue: How can the potential be realized? His answer is that experience shows that highly detailed regulatory framework can turn out to have a very negative impact while a basic, regulatory framework based on the idea of market-driven standards can help industries prosper. He said that the fourth industrial revolution is about fundamentally rethinking the way that goods are produced and services delivered. The maritime world revolves around a concept of cooperation between regulators and industry.

"Just as new possibilities for using the technology are thought of every day, the drivers behind the technology have changed. There has been a move away from thinking only in terms of safety and security, towards optimization, connectivity and human welfare on board ships," he stated [15].

The IMO has identified e-Navigation from a legal point of view as "non-mandatory instruments." In simple terms this means that e-Navigation is a good idea. The hard fact, however, is that e-Navigation models, systems, and equipment have to prove their excellence, including risk-reducing and cost-beneficial effects, in competition with mandatory instruments. As stated in Chapter 5, there is currently no intention by the IMO to mandate the carriage on board of e-Navigation equipment.

The e-Navigation Underway Conference report on the presentation by Robert Ward, president of the IHO, covered his description of how "hydrography and nautical charting will underpin many aspects of e-Navigation, which he described as the maritime "Intranet," where mariners, ship operators, and all those involved in commercial maritime activity provide digital information only once, and can obtain relevant authoritative information in return. Information will be available when it is needed; and most important, it will be able to be combined and analyzed with other information, thereby making life at sea and ashore easier, safer, more efficient, and less stressful" [16].

6.6 Skills and Training

The concept of e-Navigation is proposed to equip mariners with modern proven tools to make navigation and communications more reliable and to reduce errors. The increased use of technology onboard is inevitable—whether mandated by the IMO, voluntarily adopted by the shipowner, or even brought onboard by the mariners as a personal device.

"To me, the most important aspect of the original proposal for e-Navigation (MSC 81/23/10—December 2005) is the stated compelling need that "if such technological advancement remains uncoordinated, there is a risk that the future development of the global shipping industry will be hampered through lack of standardization on board and on land, incompatibility between vessels, and an increased and unnecessary level of complexity." This was, and remains, a considerable risk and I'm pleased that during the last ten years of concept development these issues of coordination and standardisation have been identified as priorities," says Captain David Patraiko, FNI, Director of Projects of the NI in London. The NI is an international representative body for maritime professionals worldwide.

He continues: "Over the same ten years there has been an often asked question of training and particularly will there be a need for specific competencies written into STCW and a need for a correlating IMO model course? I don't think so. If e-Navigation is implemented as envisioned, mariners will not necessarily know they are using something called e-Navigation. They will be using systems that may be e-Navigation enabled, they may use information provided in accordance with e-Navigation protocols, and they may benefit from tools that are seamlessly integrated into navigation and communication systems that are a progression of those tools that have evolved over many years. If e-Navigation correctly harmonises modern systems, mariners should not necessarily be able point to something called e-Navigation on their bridges. e-Navigation should be more of a gradual evolution rather than a revolution."

Captain Patraiko said that mariners will, however, need more education and training in managing information and working with decision support systems. This is not unique to the maritime domain. In every walk of life, people are being bombarded with greater levels of data and information aimed at improving their life somehow. On board ships, for the purpose of navigation, this information will include meteorological information, hydrographic, oceanographic, PNT, AtoN, and traffic, port operations just for starters. This is before you take into account all the information pertaining to commercial and regulatory obligations.

Mariners will need to be better at judging the quality of the information provided to them; they will need to know how much information they can use before they become overloaded, and they will need to manage information provided by communications versus information derived through their own senses such as eyes, ears, and touch. They will need to determine how advice provided by shore operators or decision support systems compare with their own judgment and experience. From a VTS point of view, they too will need to intelligently interpret greater levels and new types of information. And as communication between ship and shore becomes more technically prevalent, the human element issues of good communication and the need for best practice will also need to be addressed. This will become more evident as concepts such as STM mature in our multilingual and multicultural industry.

Patraiko provided an example: From a shipboard perspective, mariners have access to more meteorological information that ever before in the history of navigation. However, they also have access to more weather prediction algorithms and services than ever before— so how should they choose between conflicting reports? With so many PNT systems available, how do they choose? If they are told about a UKC solution based on nonship data, can they trust it? Where does the responsibility/accountability rest? Currently, ships' bridges bristle with technology, often bringing real benefits but sometime introducing some confusion. ECDIS on the bridge now allows ship's officers to monitor the ship's position in real time while under pilotage without having to run back to the chart room. The ability to have a radar continually monitor targets by cojoining ARPA and AIS data gives mariners a very powerful tool for assessing risk of collision with greater accuracy. However, Captain Patraiko underlined that the greater need for standardization to facilitate better familiarization of ship systems can be seen by the support for an S-Mode.

When it comes to shipowners and operators, they will need to ensure that the mariners they hire are capable of working in new ways. Mariners in the future will need to have greater analytical skills that come from general education rather than maritime training. Such skills can't be written into STCW competencies nor are they specific to navigation; they will apply across the ship, including engineering, cargo operations, and regulatory compliance.

Finally, e-Navigation will offer new tools and resources and that can be a good thing. If they are coordinated and standardized and the mariners are intelligent in their use, then safety and efficiency will be improved. Future technological development will bring new opportunities and risks. "We are already seeing increased use of communication and automation, and I'm sure that applications of Artificial Intelligence (AI) and Augmented Reality (AR) for maritime navigation are only just around the corner. Is there a risk that if not done correctly there can be incompatibility between vessels, and an increased and unnecessary level of complexity—the answer is yes. System engineers must keep a focus on harmonization and HCDs and users, both ship and shore, need to be recruited or educated with the analytical skills needed to use these systems and master a greater level of information management and system integration," Captain Patraiko says.

6.7 Unmanned Ships

"The benefits of automation and autonomy to varying degrees have become self-evident in many areas of commercial activity including transportation. Many cities worldwide have automated and partially autonomous rail or metro systems. Fully automated container terminals are well established as efficient ways to manage the ship-to-shore interface. Airline travel now depends in large part on automation within flight systems for certain stages of the flight program [...].

While automation and autonomy in shipping has been considered for some time with various systems already being field-tested, it is still seen as a relatively long way off and remains mired in uncertainty over a multitude of technical, practical, and regulatory concerns. That said, the continued drive for extracting cost efficiencies and increasing competitiveness within the transportation chain continues to be a powerful driver in this field" [17].

Regarding recent developments in Europe, Maritime Reporter and Engineering News reports, "The joint European Research Project MUNIN was finalized in August 2015 [...]. Participating partners in the 3-year project were research institutes, universities, and private companies from eight European countries including Germany, Norway, Sweden, Iceland, and Ireland (see Figure 6.1). The concept of autonomous shipping was analyzed both theoretically and through developing practical solutions and technical systems for autonomous operation of a ship. There were many matters under [scrutiny], from needed developments in ship-to-shore communications to fully enable autonomous operations, including legal aspects, risks, and economics of such systems. The main practical focus was directed at the development of autonomous control systems aboard a bulk carrier remotely controlled from a shore control center, simulation-based and integrating the various functions needed.

A shore control center was designed that was capable of supervising vessels worldwide using electronic nautical charts [...]. The system included a built-in possibility for outside intervention. The relating unmanned bridge development included an automated lookout system capable of conducting evasive actions autonomously and according to international regulations from preventing collisions at sea . An advanced sensor system for this purpose was also developed and tested. A system had been developed to

Figure 6.1 The MUNIN project. (Illustration © Fraunhofer CML.)

recognize small objects on the water surface and transmit warnings accordingly. A weather routing system providing an efficient and safe journey, already common on many ships today, was considered as a natural part of a future unmanned vessel. In addition, an autonomous engine Room was developed with relevant engine functions monitored remotely on multiple monitors. Remote condition monitoring was enabled to predict eventual engine service needs at an early stage, enabling servicing to be done when the vessel is in port.

With relatively low $3.5 million budget, the MUNIN project achieved the realization of a virtual system that safely analyzed and tested unmanned ship operation in a maritime environment where there were also conventional vessels in operation, a likely future scenario. Another outcome from the MUNIN project was the idea of a similar classification for an unattended watch-free bridge arrangement, to be applied during deep-sea voyages, comparable to that of the unattended machinery spaces that already exist [...].

MUNIN is just one example of an unmanned shipping development project, and there are a number of similar projects initiated on national levels and by maritime companies independently or through industry collaborations. [However], cybersecurity is an increasing concern for shipping, and will be a crucial element in unmanned ship operations [...]. The area of cybersecurity will see a lot more attention in the years to come, and will be addressed in the rules and procedures" [18].

"The topic of unmanned shipping goes beyond e-Navigation, but research in this area does profit greatly from e-Navigation," according to Hans-Christoph Burmeister from Fraunhofer CML in Germany, who also mentions goals such as improved reliability, stability, and resilience of navigational navigation. "Thanks to on-site testing and simulations of navigational sensor fusion, MUNIN is providing important input in this field," he says [19].

6.8 Big Data

As e-Navigation evolves from concept to operational reality, the importance of harmonization of systems and equipment and integration and standardization is one of crucial factors to its success.

Using information technology in combination with communication opportunities places e-Navigation in a central role in a bigger and wider picture.

The maritime world is, as many other sectors, a part of Industry 4.0—the fourth technological revolution in which computers and automation will come together in an entirely new way, with robotics connected remotely to computer systems equipped with machine-learning algorithms that can learn and control the robotics with very little input from human operators.

According to an article in *Forbes* [15] written by Bernard Marr, "Industry 4.0 introduces what has been called the 'smart factory,' in which cyberphysical systems monitor the physical processes of the factory and make decentralized decisions. The physical systems become an IoT, communicating and cooperating both with each other and with humans in real time via the wireless web" [20]. Industry 4.0 is the forth milestone in a development, in which "first came steam and the first machines that mechanized some of the work our ancestors did. Next was electricity, the assembly line, and the birth of mass production. The third era of industry came about with the advent of computers and the beginnings of automation, when robots and machines began to replace human workers on those assembly lines" [20].

Shipping of the future is closely linked to ICT with the possibility of being online everywhere all the time. The potential for combining this data is enormous. Andreas Nordslett, director-general of the Danish Maritime Authority, says that smart, competitive shipping, ready for the future, depends on us unlocking this potential. To realize this potential, each stakeholder has a role to play and a coordinated approach will be needed for the work ahead.

"Legislators must provide the right, legislative framework to support innovation, development and up take. The process can only be market-driven. The industry side must be actively involved and committed for the potential to be realized. The key to unlocking this potential for shipping of the future is a coordinated approach, which means pulling in the same direction across authorities, business sectors, regions and international fora. The coordinated approach must rest on sensible, smart solutions to support smart shipping. Smart, competitive and sustainable shipping depends, first of all, on efficiency. And to be sharp and efficient today and in the future shipping must embrace and fully use the potential of ICT-solutions," he emphasizes [15].

The director-general says that this big picture is bigger than e-Navigation. It is about ICT changing the way we do business every day. Maritime data is collected around the clock by a myriad of actors around the world. e-Navigation is one of the ways in which information technology can be embraced. big data is already out there on the oceans, waiting to be fully utilized. Provided the right infrastructure is put in place, the maritime community will end up with a box of new and exciting tools [15],

Another view comes from a dissertation submitted to the World Maritime University in Sweden, "Major Challenges and Solutions for Utilizing Big Data in the Maritime Industry," by Sadaharu Koga from Japan in 2015, compared e-Navigation and big data (see example in Figure 6.2) [21]:

> As regards e-Navigation, there is both a similarity and a difference when compared with the notion of big data. The similarity was that both e-Navigation and maritime big data totally depended on electronic devices and the data derived and processed by them. The difference was that e-Navigation concentrates on safety and efficiency of navigation during voyages and is interested in real-time, streaming data, while maritime big data covers not only real-time data, but also the accumulation of data for use after voyages.

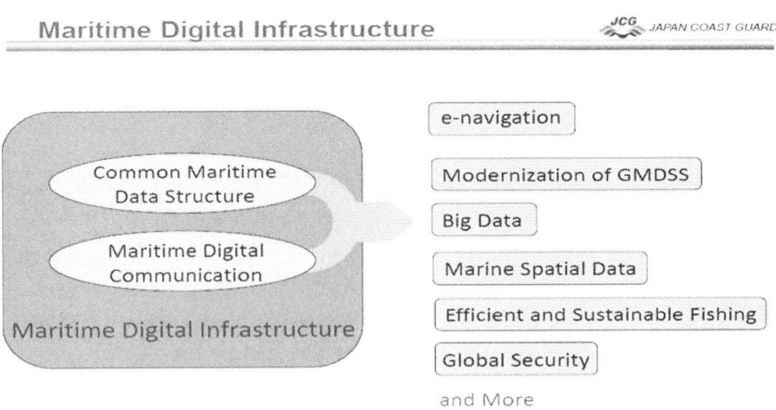

Figure 6.2 Maritime digital infrastructure as presented by the Japan Coast Guard [22].

Taking into account this similarity, the direction of development and tasks of e-Navigation were reviewed and it was concluded that there are four tasks that may be linked with maritime big data. The four tasks are automated data collection, software quality assurance, reliability and resilience of data, and CMDS.

The issue of technology is a crucial foundation of development of maritime big data. However, according to Sadaharu Koga, this issue should be tackled by the private sector depending on the principle of market competitiveness. Therefore, the key was how it should be supported by governments or other third parties. in terms of such support, two main approaches were considered: flexibility and data integration. Regarding flexibility, the government is supposed to provide test bed or corresponding budgetary aid for innovative projects through research and development projects. This kind of support will give incentives for private companies and institutions to venture to materialize their innovative ideas. In terms of data integration, unification of format or structure of data may be helpful for the maritime industry to exchange their data smoothly. This could be achieved in the IMO by reference to a CMDS framework in the discussion of e-Navigation. However, the IMO member states have to be careful to ensure the scalability of format or structure of data. Too rigid standards sometimes prevent further innovations. When unification of the data style is carried out, the classification societies can play a supplementary role providing data certification services, the dissertation of Sadaharu Koga proposes [21].

6.9 Managing the Environmental Impact of Shipping

The potential for climate change due to increased GHG concentrations in the atmosphere is quite high and transport through ports and port operations are contributors. Additionally, stricter air quality standards and concern for nearby residents prompt port authorities to take stricter measures to improve air quality in their ports.

"The world's key ports have committed themselves to reducing GHG emissions while continuing their role as transportation and economic centers. This commitment is called the World Ports Climate Initiative (WPCI). This initiative seeks to implement the principle of sustainability in supply chains, taking into account local circumstances and varying port management structures. The initiative also supports seagoing ships taking measures to reduce ship-to-air GHG emissions.

One of the projects within WPCI is the development of an environmental ship index (ESI). The ESI identifies seagoing ships that perform well in reducing more air emissions than required by the current emission standards of the IMO. The ESI evaluates the amount of nitrogen oxide (NO_X), and sulfur oxide (SO_X) that are released by a ship measures the increase in the efficiency of ships regarding their fuel consumption, (with a view to CO2 reduction) and the reduction achieved in the release to the atmosphere of GHG. The ESI is a good indication of the environmental performance of ocean-going vessels and will assist in identifying cleaner ships in a general way" [23].

The index is intended to be used by ports to reward ships when they participate in the ESI and will promote clean ships, but can also be used by shippers and ship owners as their own promotional instrument.

At the same time, environmentally friendly ships are being built, works that are supported by e-Navigation. Solar Frontier's CIS thin-film solar panels have been installed by Kawasaki Kisen Kaisha, Ltd. ("K"Line), one of Japan's largest shipping companies. CIS stands for copper (C), indium (I), and selenium (S), and the technology is very different from that of the conventional crystalline silicon solar panels. CIS offers advantages in terms of power generation, efficiency, and environmental impacts [24].

"K" Line's ship, a car-carrier named *Drive Green Highway,* was the third of ten similar vessels that will be built under "K" Line's "Drive Green Project" (Figure 6.3). *Drive Green Highway* uses its sheer size to improve its emissions efficiency. The vessel is 200 meters long, 37.5 meters wide, and uses an improved design that enables it to carry up to 7,550 passenger vehicles at once [25].

"Drive Green Highway integrates the world's most advanced environmentally friendly and energy-saving technologies, which range from its GHG suppression systems in its engines to Solar Frontier's CIS solar panels on its decks.

Combining the advantages of its design and technology improvements, *Drive Green Highway* will emit 25% less carbon dioxide

Figure 6.3 *Drive Green Highway* integrates advanced energy-efficient technologies and design improvements to reduce its impact on the environment. (Photo: "K" Line.)

(CO2), 50% less nitrogen oxide (NOx), and 90% less sulfur oxide (SOx) per vehicle transported.

Contributing to *Drive Green Highway*'s energy efficiency is one of the largest solar energy systems on any ship in the world. More than 900 of Solar Frontier's CIS solar panels have been installed on its top deck, totaling 150 kilowatt-peak of electricity generating capacity. The electricity generated by these solar panels is used to power all LED lighting on the vehicle deck.

'K Line is demonstrating how solar energy can improve the energy efficiency and reduce the ecological impact of the shipping industry,' says Atsuhiko Hirano, CEO of Solar Frontier, the supplier of the CIS panels. 'This is one example of the greater role that solar energy has to play in our future, supporting an ever broader range of technologies in a wider range of industries'" [25].

Eidesvik Offshore in Norway is a supply, subsea, and seismic shipping company. The fleet consists of supply ships, supply vessels, subsea vessels, and seismic vessels.

For Eidesvik Offshore the focus on innovation is largely about the environment. In 2013, the shipping company celebrated the 10th anniversary of their first LNG ship, *Viking Energy*. They are now the world's largest operator of this type of vessel. By using LNG as fuel reduces emissions of soot, NOx, CO_2, and sulfur significantly, according to the shipping company.

Also in 2013, Eidesvik Offshore was the first company in the world to use battery technology on board an offshore vessel, which is an efficient and very environmentally friendly way to operate a ship. The success of *Viking Energy* was motivating, and in 2009 it built *Viking Lady* (Figure 6.4). This advanced vessel has fuel cells in addition to dual fuel engines on board, which will benefit both LNG and diesel. The ship received great attention at the UN Climate Change Conference that year. In 2013, the boundaries were pushed by installing a battery pack on board.

In 2016, Eidesvik Offshore installed a battery package onboard *Viking Energy* and obtained, as the first vessel in the world, the class notation Battery Power. The vessel is the first real hybrid of its class.

However, the situation in the global oil and gas industry has changed significantly. The steep decline in oil prices has forced the entire industry to make significant changes in their operations.

Figure 6.4 Viking Lady. (Photo: Eidesvik Offshore ASA.)

An interesting aspect in this respect is if the current market situation has worked as a deterrent toward further investments and focused the direction of further environmental and technological innovation.

"Even in these difficult times, we have made agreements with clients for installation of battery packages, however low oil prices could slow down the green shift. The oil companies charter in the vessels for which they obtain lowest rates, and those vessels are normally the oldest and not very environmentally friendly," says Jan Fredrik Meling, president and chief executive officer (CEO).

Another question during these challenging times in the offshore shipping industry is: How important is it to offer vessels with high environmental and technological benefits to the industry?

"The real issue is to what extent you can demonstrate lower consumption of fuel, and thereby contribute to cost reduction for your client. The environmental effect of lower consumption comes as and addition, but cost is what matters these days, for our customers as well as for ourselves," CEO Meling says.

e-Navigation with its improved connectivity—allowing ships to optimize route planning and time at sea and will save fuel, reduce emissions, and improve navigation—will also contribute to reducing accidents and their consequential oil spills and damage to the environment. This is in line with the examples above and the need by owners to reduce costs, reduce risks, and protect the environment.

Will the benefits of e-Navigation be a cost-effective solution for the owners or will the pressure to increase margins in a tight market cause a lack of investment in modern solutions such as e-Navigation? More has to be done to sell the benefits of e-Navigation in the long term to shipowners. A way has to be found to cut through the restrictions of the existing ship's equipment regulations so the shipowner has more freedom to install what is needed to benefit his or her business regardless of regulations made several years ago. An example for further thought: Aircraft pilots use iPads on the flight deck, so why not ships' bridge teams? In 2011, United Continental Holdings, Inc. announced that it was "converting to paperless flight decks and deploying 11,000 iPads to all United and Continental pilots. The electronic flight bags (EFB) replace paper flight manuals, and as a first for major network carriers, provide pilots with paperless aeronautical navigational charts through an iPad app [...]. 'The paperless flight deck represents the next generation of flying,' says Captain Fred Abbott, United's senior vice president of flight operations. 'The introduction of iPads ensures our pilots have essential and real-time information at their fingertips at all times throughout the flight'" [26].

6.10 Presenting the Future

In his first speech after assuming office in 2016, the IMO secretary general, Mr. Kitack Lim, said, as stated in Chapter 1, that the challenge was to turn the future into the present, so that the much-heralded benefits of e-Navigation could be fully realized.

So far, the IMO has developed a strategy and an associated plan to implement its strategy and is currently coordinating the development of rules and guidance for the introduction of new technology. The IMO-led evolution of e-Navigation will likely go down into history as the international maritime community's regulatory response to harmonize the traditionally incremental introduction of new technology onboard ships and ashore, particularly in the areas of marine navigation and communications and including a recognition of the important role of the human element.

While uncoordinated introduction of technology and its impact on the human element was a key trigger, the IMO's work on e-Navigation retained a focus on user needs throughout. A key user need identified was that new systems should be more useable. In this regard, the IMO should be recognized for its first e-Navigation "product," which is a guideline on software quality assurance and human-centered design.

There is perennial tension between the quests for profit in a capital-intensive industry like shipping and the cost of training officers. This is not a new issue—it has prevailed for decades. New training requirements for e-Navigation (such as training for new or more integrated systems), in parallel with mandatory training required by STCW, may lead to an undesirable situation in which crews will not have the level of competency required to make the best use of increasingly more complex navigation systems.

The introduction of increased automation on board ships, and the resultant and inevitable reduction in human engagement, could introduce new risks such as a lack of mode awareness. Personnel may need to undergo extra training to fully understand the automated systems designed to support navigation—and vitally, to know what to do if things go wrong. For example, the Air France flight AF 447 that crashed into the Atlantic Ocean in 2009, killing everyone on board was caused by human and techical errors. This sowed the seed for a complex error chain that ended tragically [27].

In the future, mariners will benefit from better ergonomic design of ships bridges. Human-centered design principles will ensure that navigation systems and bridges in general are designed and built to better suit user needs, not the other way around. Bespoke information is likely to be provided to officers when they need it via maritime applications or apps, just like new-generation seafarers, as described in the introduction to this chapter, are used to in their lives ashore. Ships will increasingly become cyber-enabled maritime segments within an increasingly interconnected and digitalized global supply chain. The development of autonomously operated vessels will no doubt continue.

In the future, there needs to be an internationally agreed-on common data structure to optimize the exchange of relevant information and data in the maritime domain. The objective of the joint IMO/IHO harmonization work on data modeling, is to harmonize efforts with the goal of creating a robust and extendable data structure. Such measures will harmonize the implementation of e-Navigation onboard, and go hand in hand with implementation processes ashore.

As the global standard-setting authority for maritime safety, security, and environmental protection, as much as possible the IMO maintains a regulatory framework that is equitable, effective, and universally adopted. International rules and regulations can, at times, overlap, or be seen as burdensome or outdated. Over-regulation can affect adversely on national administrations and industry and threaten the sustainability of maritime transport. The successful implementation of e-Navigation will depend on technical harmonization and standardization on a global scale, an effective regulatory framework, and cooperation between all maritime stakeholders.

In closing, while work at the IMO and other organizations is focused on several deliverables, patchy implementation ashore is foreseen. Some authorities will do much better than others. However, on board, IMO rules such as existing carriage requirements for navigation equipment and their relevant performance standards will define what can and cannot be done. A way forward in this respect could be to harmonize some of the existing performance standards and guidelines and to separate them from technology. This will open the door for innovative, cost-effective systems and equipment, which will further contribute to the reduction of collisions and groundings.

One thing remains certain, technology will continue to be developed and used to find ways to reduce costs, maximize profits, and, somewhere in the mix, help improve safety This will necessitate an ongoing effort by all concerned. Looking to the future, the absence of globally harmonized solutions will give rise to the real risk that regional solutions will be implemented. There are already signs of this occurring in some parts of the world. Such a trend will lead to confusion and could be detrimental to maritime safety. In addition, the reputation of international rule-making and coordinating bodies like the IMO, IHO, and IALA may be tarnished. The platform for the establishment of the e-Navigation concept is identification of user needs, followed by international agreement of implementation of several tasks to capture those needs. Interaction between technological systems and the human element are becoming more complex and require a better understanding of how to enhance safety, security, and efficiency by harmonization of marine navigation systems and equipment and supporting shore services. This is the core of e-Navigation and when implemented it will improve the maritime environment, not just for the benefit and safety of all mariners, but for the global community and planet as a whole.

References

[1] Tapscott, D., *Grownup Digital: How the Net Generation is Changing Your World*, New York: McGraw-Hill, 2008.

[2] McKinsey & Company, *Digital Globalization: The New Era of Global Flows*, March 2006.

[3] De Bièvre, A., "Automated Ship Reporting Gains IMO Support," *Marine Electronics and Communications*, June 17, 2016, http://www.marinemec.com/news/view,automated-ship-reporting-gains-imo-support_43382.htm, accessed May 10, 2017.

[4] International Chamber of Shipping, *Shipping and World Trade Overview*, available at http://www.ics-shipping.org/shipping-facts/shipping-and-world-trade,accessed May 10, 2017.

[5] Wingrove, M., *Shipping Needs to Cultivate Disruptive Technology*, March 2, 2016, available at http://www.containerst.com/news/view,shipping-needs-to-cultivate-disruptive-technology_42072.htm, accessed May 10, 2017.

[6] Marine Electronics and Communication, "Disruptive Digital Technology Will Solve Shipping Issues," April 21, 2016, available at http://www.marinemec.com/news/view,disruptive-digital-technology-will-solve-shipping-issues_42670.htm, accessed May 10, 2017.

[7] IMO, Document C 116/4/1.

[8] DNV GL, *Technology Outlook 2025*, available, at http://to2025.dnvgl.com/shipping/digitalization/, accessed May 10, 2017.

[9] Gahnström, J., "e-Navigation from the End User's Perspective," presented at *e-Navigation Underway 2017*, available at https://www.intertanko.com/Global/presentations/2017/E_Navigation_underway_2017.pdf, accessed May 10, 2017.

[10] Zachariae, F., Presentation at the *International Habour Masters Association (IHMA) Conference*.

[11] IMO, *Trends, Developments and Challenges Facing the IMO in the 2018–2023 Period*, Document C 116/4/1.

[12] Richardson, K., "e-Navigation…the Implications for Ports," A Presentation for the *e-Navigation Underway Conference*, Copenhagen/Oslo, January 31–February 2, 2017.

[13] IMO, *Guideline on Software Quality Assurance and Human-Centred Design for e-Navigation*, Document MSC.1/Circ.1512, July 13, 2015.

[14] Mahapatra, A., "The Global Approach to Automated and Standardized Ship Reporting," presented at *Straits e-Navigation Alliance Conference*, 2015.

[15] Nordslett, A., *Conference Report e-Navigation Underway*, 2016.

[16] IALA AISM Conference Report e-Navigation Underway 2015, available at http://www.iala-aism.org/content/uploads/2016/06/e-Navigation-Underway-2015-Report-20150202.pdf, accessed May 10, 2017.

[17] Norwegian Hull Club, *Autonomous Shipping Concepts*, available at https://www.norclub.no/blog/autonomous-shipping-concepts/, accessed May 10, 2017.

[18] Segercrantz, H., *Unmanned Vessels: The Future is Now*, October 26, 2015, available at http://maritimeglobalnews.com/news/unmanned-vessels-future-bxcnzh, accessed May 10, 2017.

[19] Fraunhofer Center for Maritime Logistics, press release, *Unmanned Shipping: a Topic at e-Navigation Underway 2014*, February 12, 2014, available at http://www.cml.fraunhofer.de/content/dam/cml/en/documents/PressReleases/140211%20PM%20e-Navigation%20Fraunhofer%20CML%20EN.pdf, accessed May 10, 2017.

[20] Marr, B., "What Everyone Must Know About Industry 4.0," *Forbes Magazine*, June 20, 2016, available at https://www.forbes.com/sites/bernardmarr/2016/06/20/what-everyone-must-know-about-industry-4-0/#39ee3121795f, accessed May 10, 2017.

[21] Koga, S., "Major Challenges and Solutions for Utilizing Big Data in the Maritime Industry," *The Maritime Commons: Digital Repository of the World*, Maritime University, 2015, available at http://commons.wmu.se/cgi/viewcontent.cgi?article=1489&context=all_dissertations, accessed May, 2017.

[22] Noguchi, H., *e-Navigation and Maritime Digital Infrastructure*, January, 26, 2016, available at https://www.iho.int/mtg_docs/com_wg/MSDIWG/MSDIWG7/Open/OF13_Hideki%20Noguchi%20(Vice-Chair%20of%20IALA%20e-Navigation%20Committee).pdf, accessed May 10, 2017.

[23] Paris Process on Mobility and Climate (PPMC), *Environmental Ship Index Part of the International Association of Ports & Harbors' World Ports Climate Initiative*, available at http://www.ppmc-transport.org/environmental-ship-index-part-of-the-international-association-of-ports-harbors-world-ports-climate-initiative/, accessed May 10, 2017.

[24] Solar Frontier, What is CIS? Available at https://www.solar-frontier.eu/en/cis-technology/what-is-cis/, accessed May 10, 2017.

[25] News Hound, "Eco-Ship Drive Green Highway Launched by K-Line," on *International Institute of Marine Surveying* website, available at https://www.iims.org.uk/eco-ship-drive-green-highway-launched-by-k-line/, accessed May 10, 2017.

[26] *United Continental Holdings, Inc. United Airlines Launches Paperless Flight Deck with iPad*, August 23, 2011, available at http://www.prnewswire.com/news-releases/united-airlines-launches-paperless-flight-deck-with-ipad-128240343.html, accessed May 10, 2017.

[27] Wise, J., "Air France 447 and the Limits of Aviation Safety," July 9, 2012, available at http://www.popularmechanics.com/flight/a7813/air-france-447-and-the-limits-of-aviation-safety-10487501/, accessed May 10, 2017.

About the Author

John Erik Hagen is regional director at the Norwegian Coastal Administration.

He pursued a career in the Norwegian Navy, attending Officer Training School and later the Norwegian Defence University College. He has also a civil university degree in law, social science, and business administration.

After his naval career, he spent several years in managing positions in the offshore industry and transportation services. Over a period, he functioned as acting director-general for the Norwegian Coastal Administration.

He has served as coordinator of the IMO Correspondence Group on E-navigation, and as chairman of the IMO working groups on e-Navigation. He was also chairman of the High Level Advisory Board on its work on The Straits of Malacca and Singapore: Efficient, Safe and Secure Ship Traffic Management System (the SESAME project). John Erik Hagen has also been the Norwegian delegation's chairman of the Working Group established for the cooperation between the Russian Federation and Norway on maritime safety in the Barents area.

Index

M

N

O

The GNSS Technology and Applications Series

Elliott Kaplan and Christopher Hegarty, Series Editors

A-GPS: Assisted GPS, GNSS, and SBAS, Frank van Diggelen

Applied Satellite Navigation Using GPS, GALILEO, and Augmentation Systems, Ramjee Prasad and Marina Ruggieri

Digital Terrain Modeling: Acquisition, Manipulation, and Applications, Naser El-Sheimy, Caterina Valeo, and Ayman Habib

Geographical Information Systems Demystified, Stephen R. Galati

GNSS Applications and Methods, Scott Gleason and Demoz Gebre-Egziabher

GNSS Interference Threats and Countermeasures, Fabio Dovis, editor

GNSS Markets and Applications, Len Jacobson

GNSS Receivers for Weak Signals, Nesreen I. Ziedan

GNSS for Vehicle Control, David M. Bevly and Stewart Cobb

GPS/GNSS Antennas, B. Rama Rao, W. Kunysz, R. Fante, and K. McDonald

Implementing e-Navigation, John Erik Hagen

Introduction to GPS: The Global Positioning System, Second Edition, Ahmed El-Rabbany

MEMS-Based Integrated Navigation, Priyanka Aggarwal, Zainab Syed, Aboelmagd Noureldin, and Naser El-Sheimy

Navigation Signal Processing for GNSS Software Receivers, Thomas Pany

Principles of GNSS, Inertial, and Multisensor Integrated Navigation Systems, Second Edition, Paul D. Groves

RF Positioning: Fundamentals, Applications, and Tools, Rafael Saraiva Campos, and Lisandro Lovisolo

Server-Side GPS and Assisted-GPS in Java™, Neil Harper

Spread Spectrum Systems for GNSS and Wireless Communications, Jack K. Holmes

Understanding GPS/GNSS: Principles and Applications, Third Edition, Elliott Kaplan and Christopher Hegarty, editors

Ubiquitous Positioning, Robin Mannings

Wireless Positioning Technologies and Applications, Second Edition, Alan Bensky